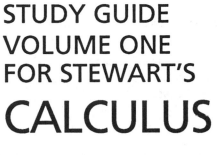

STUDY GUIDE
VOLUME ONE
FOR STEWART'S
CALCULUS

THIRD EDITION

RICHARD ST. ANDRE
Central Michigan University

BROOKS/COLE PUBLISHING COMPANY
I**T**P An International Thomson Publishing Company

Pacific Grove • Albany • Bonn • Boston • Cincinnati • Detroit • London • Madrid • Melbourne
Mexico City • New York • Paris • San Francisco • Singapore • Tokyo • Toronto • Washington

Sponsoring Editor: Elizabeth Rammel
Editorial Assistant: Carol Benedict
Production Coordinator: Dorothy Bell
Cover Design: Vernon T. Boes
Cover Sculpture: Christian Haase
Cover Photo: Ed Young
Printing and Binding: Malloy Lithographing, Inc.

For more information, contact:

BROOKS/COLE PUBLISHING COMPANY
511 Forest Lodge Rd.
Pacific Grove, CA 93950
USA

International Thomson Editores
Campos Eliseos 385, Piso 7
Col. Polanco
11560 México D. F. México

International Thomson Publishing Europe
Berkshire House 168-173
High Holborn
London WC1V 7AA
England

International Thomson Publishing GmbH
Königwinterer Strasse 418
53227 Bonn
Germany

Thomas Nelson Australia
102 Dodds Street
South Melbourne, 3205
Victoria, Australia

International Thomson Publishing Asia
221 Henderson Road
#05–10 Henderson Building
Singapore 0315

Nelson Canada
1120 Birchmount Road
Scarborough, Ontario
Canada M1K 5G4

International Thomson Publishing Japan
Hirakawacho Kyowa Building, 3F
2-2-1 Hirakawacho
Chiyoda-ku, Tokyo 102
Japan

Printed in the United States of America

10 9 8 7 6 5

ISBN 0-534-21804-0

Preface

This Study Guide is designed to supplement the first eleven chapters of Calculus, 3rd edition, by James Stewart. It may also be used with Single Variable Calculus, 3rd edition. If you later go on to multivariable calculus, you will want to obtain volume II of this Study Guide.

Your text is well written in a very complete and patient style. This Study Guide is not intended to replace it. You should read the relevant sections of the text and work problems, lots of problems. Calculus is not a spectator sport.

This Study Guide captures the main points and formulas of each section and provides short, pithy questions that will help you understand the essential concepts. Every question has an explained answer. The pages are perforated so that you can detach them. The two column format allows you to cover the answer portion of a question while working on it and then uncover the given answer to check your solution. Working in this fashion leads to greater success than simply perusing the solutions.

As a quick check of your understanding of a section work the page of On Your Own questions located toward the back of the Study Guide. These are all multiple choice questions -- the type that you might see on an exam in a large-sized calculus class. You are "on your own" in the sense that an answer, but no solution, is provided for each question.

I hope that you find this Study Guide helpful in understanding the concepts and solving the exercises in Calculus, 3rd edition.

Richard St. Andre

Contents

Contents (Continued)

Contents (Continued)

Contents (Continued)

Please cut page down center line.
Use the half page to cover the right
column while you work in the left.

Review and Preview

Cartoons courtesy of Sidney Harris. Used by permission.

Functions and Their Graphs

Concepts to Master

A. Definition of a function; Domain; Range; Independent variable; Dependent variable

B. Implied domain

C. Graph of a function

D. Piecewise Defined Function

E. Symmetry (even and odd functions)

F. Combining Functions (sum, difference, product, quotient, and composition)

Summary and Focus Questions

A. A function f is a rule that associates pairs of numbers: to each number x in a set (the <u>domain</u>) there is associated another real number, denoted $f(x)$, the image of x. The set of all images ($f(x)$ values) is the <u>range of f</u>. The variable x is the <u>independent</u> variable and $y = f(x)$ is the dependent variable.

For example, the function S, "to each nonnegative number, associate its square root," has domain $[0, \infty)$ and rule of association given by:

$$S(x) = \sqrt{x}.$$

1) Sometimes, Always, Never:

Both of the points $(2, 5)$ and $(2, 7)$ could be pairs of a function f.

Never. $f(2)$ cannot be both 5 and 7.

2) True or False:

$x = y^2$ defines y as a function of x.

False, since $(4, 2)$ and $(4, -2)$ satisfy the equation. Note that x is a function of y.

3) A track is in the shape of two semicircular ends and straight sides as in the figure. Find the distance around the track as a function of the radius.

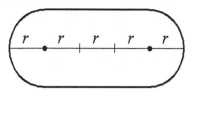

Each straight side is $3r$ and each semicircular end is πr. The distance d is $d = 2(3r + \pi r)$ or $d = (6 + 2\pi)r$.

B. The domain of a function f, if unspecified, is understood to be the largest set of the real numbers x for which $f(x)$ exists. Thus $f(x) = \dfrac{1}{\sqrt{x-5}}$ has implied domain $(5, \infty)$. The range of $f(x)$ is $(0, \infty)$.

4) Find the domain of $f(x) = \sqrt{x^2 - 16}$.

For $f(x)$ to exist, $x^2 - 16 \geq 0$. Thus $(x-4)(x+4) \geq 0$. The solution set is $x \leq -4$ or $x \geq 4$. The domain is $(-\infty, -4] \cup [4, \infty)$.

5) Find the domain of $f(x) = \dfrac{1}{|x|+1}$.

Since $|x| \geq 0$ for all x, $|x| + 1$ is never 0. The domain is all real numbers.

C. The <u>graph</u> of $y = f(x)$ is all points (x, y) in the Cartesian plane that make $y = f(x)$ true. About all we can do now to sketch graphs is plot points and recognize certain graphs (such as lines or parabolas) from the form of the equation.

6) Sketch the graph of
$f(x) = \sqrt{x^2 - 16}$. (See question 4.)

Here are several computed values: $(4, 0)$, $(5, 3)$, $(6, \sqrt{20})$, $(-4, 0)$, $(-5, 3)$, ...
Squaring both sides of $\sqrt{x^2 - 16}$ gives $y^2 = x^2 - 16$, or $x^2 - y^2 = 16$. The graph is the top half of a hyperbola.

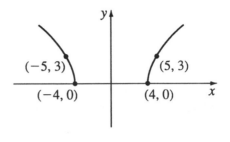

D. A piecewise defined function partitions its domain into parts and uses different formulas on each part. For example:

$$f(x) = \begin{cases} \frac{1}{x} & \text{if } x \geq 1 \\ 3x & \text{if } x < 1 \end{cases}$$

has a domain of all real numbers partitioned into $(-\infty, 1)$ and $[1, \infty)$. Use the appropriate rule to draw the graph over each part.

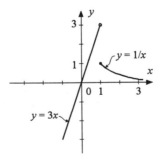

7) The phone company charges a flat rate for local calls of $12.60 per month with an additional charge of $0.05 per call after 55 calls. Express the monthly bill as a function of the number of local calls.

Let $x =$ the number of local calls. Then
$$M(x) = \begin{cases} 12.60, \text{ if } x \in [0,\ 55] \\ 12.60 + 0.05(x - 55), \text{ if } x \in (55,\ \infty) \end{cases}$$

E. Sometimes one part of a graph of a function will be a reflected image of another part:

$y = f(x)$ is <u>even</u> means $f(x) = f(-x)$ for all x in the domain. An even function is symmetric about the y-axis.

$y = f(x)$ is <u>odd</u> means $f(x) = -f(-x)$ for all x in the domain. An odd function is symmetric with respect to the origin.

8) If $(7, 3)$ is on the graph of an odd function, what other point must also be on the graph?

$(-7,\ -3)$. This is because
$3 = f(7) = -f(-7)$,
So $f(-7) = -3$.

9) Is $f(x) = 10 - 2x$ even or odd?

$f(x) = 10 - 2x$
$f(x) = 10 - 2(-x) = 10 + 2x$
$f(x)$ is <u>not even</u> since $f(x) \neq f(-x)$.
$-f(-x) = -(10 + 2x) = -10 - 2x$
$f(x)$ is <u>not odd</u> since $f(x) \neq -f(-x)$.

F. Simple functions may be combined to produce more complex combinations. Later, when principles of calculus are developed for simple functions, those priniciples may be applied to complex functions as well by applying the principles to the complex function's components.

Given two functions $f(x)$ and $g(x)$:

1) The <u>sum</u> $f + g$ associates to each x the number $f(x) + g(x)$.
2) The <u>difference</u> $f - g$ associates to each x the number $f(x) - g(x)$.
3) The <u>product</u> fg associates to each x the number $f(x) \cdot g(x)$.

4) The <u>quotient</u> $\frac{f}{g}$ associates to each x the number $\frac{f(x)}{g(x)}$, provided that $g(x) \neq 0$.

For example, the sum, difference, product, and quotient of $f(x) = x^2$ and $g(x) = 2x - 8$ are:

$$(f + g)(x) = x^2 + 2x - 8$$
$$(f - g)(x) = x^2 - 2x + 8$$
$$fg(x) = x^2(2x - 8)$$
$$\frac{f}{g}(x) = \frac{x^2}{2x-8}, \text{ for } x \neq 4.$$

The function $h(x) = \frac{x^2(x+1)}{2+x}$ may be written as $h = \frac{fg}{k}$ where $f(x) = x^2$, $g(x) = x + 1$, and $k(x) = 2 + x$.

For $f(x)$ and $g(x)$, the <u>composite</u> $f \circ g$ is the function that associates to each x the same number that f associates to $g(x)$, that is:

$$(f \circ g)(x) = f(g(x)).$$

To compute $f \circ g(x)$, first compute $g(x)$, then compute f of that result.

For example, if $f(x) = x^2 + 1$ and $g(x) = 3x$, then for $x = 5$:

$$(f \circ g)(5) = f(g(5)) = f(3(5)) = f(15) = 15^2 + 1 = 226.$$

Finally, to determine the components f and g of a composite function, $h(x) = (f \circ g)(x)$, requires an examination of which function is applied first.

For example, for $h(x) = \sqrt{x + 5}$, we may use $g(x) = x + 5$ (it is applied first, before the square root) and $f(x) = \sqrt{x}$. Then

$$h(x) = \sqrt{x + 5} = \sqrt{g(x)} = f(g(x)).$$

10) Find $f + g$ and $\frac{f}{g}$ for $f(x) = 8 + x$, $g(x) = \sqrt{x}$. What is the domain of each?

$(f + g)(x) = 8 + x + \sqrt{x}$ with domain $= [0, \infty)$. $\frac{f}{g}(x) = \frac{8+x}{\sqrt{x}}$ with domain $= (0, \infty]$.

11) Write $f(x) = \frac{2\sqrt{x+1}}{x^2(x+3)}$ as a combination of simpler functions.

There are many answers to this question. One is $f = \frac{hk}{mn}$ where $h(x) = 2$,

$k(x) = \sqrt{x + 1}$, $m(x) = x^2$, $n(x) = x + 3$.

12) Find $f \circ g$ where:

a) $f(x) = x - 2x^2$ and $g(x) = x + 1$

$f \circ g(x) = f(g(x)) = f(x + 1)$
$= (x + 1) - 2(x + 1)^2.$

b) $f(x) = \sin x$ and $g(x) = \sin x$

$f \circ g(x) = f(\sin x) = \sin(\sin x).$
This is not the same as $\sin^2(x)$.

13) Rewrite $h(x) = (x^2 + 3x)^{2/3}$ as a composition, $f \circ g$.

Let $f(x) = x^{2/3}$ and $g(x) = x^2 + 3$.
$f \circ g(x) = f(g(x)) = f(x^2 + 3x)$
$= (x^2 + 3x)^{2/3}.$
Thus $h = f \circ g$.

Types of Functions; Shifting and Scaling

Concepts to Master

A. Constant, power, polynomial, rational, algebraic, and transcendental functions
B. Horizontal and vertical shifting and stretching of functions; Reflecting

Summary and Focus Questions

A. Some of the types of functions that occur in calculus are included in this table.

Type	Description	Examples
Constant	$f(x) = c$	$f(x) = 5$ $f(x) = -\pi$
Power	$f(x) = x^a$ $a = -1, 1, 2, 3, 4, \ldots$	$f(x) = x^7$ $f(x) = x^{-1}$
Root	$f(x) = x^a$ where $a = \frac{1}{2}, \frac{1}{3}, \frac{1}{4}, \ldots$	$f(x) = \sqrt[3]{x},$ $f(x) = x^{1/10}$
Polynomial	$f(x) = a_n x^n + \ldots + a_1 x + a_0$ (n is the degree)	$f(x) = 6x^2 + 3x + 1$ $f(x) = -5x^{12} + 7x$
Rational	$f(x) = \frac{P(x)}{Q(x)}$, where P, Q are polynomials	$f(x) = \frac{7x+6}{x^2+10}$
Algebraic	$f(x)$ is obtained from polynomials, using algebra $(+, -, \cdot, /,$ roots, composition)	$f(x) = \sqrt{16 - x^2}$ $f(x) = \sqrt[3]{\frac{x^3}{x+1}}$
Transcendental	These are nonalgebraic functions and include, among others: Exponential: $f(x) = a^x$ (for $a > 0$) Logarithmic: $f(x) = \log_a(x)$ (for $a > 0$) Trigonometic: sine, cosine, tangent, ...	$f(x) = 2^x$ $f(x) = \log_4(x + 1)$ $f(x) = \cos 3x$

1) True or False:
   $f(x) = x^3 + 6x + x^{-1}$ is a
   polynomial.

   False, (because of the x^{-1} term).

2) True or False:
   $f(x) = x + \frac{1}{x}$ is rational.

   True. $\left(x + \frac{1}{x} = \frac{x^2+1}{x}\right)$

3) True or False:
   Every rational function is an algebraic function.

   True.

4) $f(x) = 6x^{10} + 12x^4 + x^{11}$ has degree _____.

   11.

5) True or False:
   $f(x) = 2^\pi$ is an exponential function.

   False. 2^π is constant (approximately 8.825).

B. Adding or subtracting a constant to either the dependent or the independent variable shifts the graph up, down, left, or right.

For $c > 0$:

Function	Effect on graph of $y = f(x)$
$y = f(x - c)$	shift *right* by c units
$y = f(x + c)$	shift *left* by c units
$y = f(x) + c$	shift *upward* by c units
$y = f(x) - c$	shift *downward* by c units

Multiplying either the dependent or the independent variable by a nonzero constant will stretch or compress a graph.

For $c > 1$:

Function	Effect on graph of $y = f(x)$
$y = cf(x)$	Stretch *vertically* by a factor of c.
$y = \frac{1}{c}f(x)$	Compress *vertically* by a factor of c.
$y = f(cx)$	Compress *horizontally* by a factor of c.
$y = f\left(\frac{x}{c}\right)$	Stretch *horizontally* by a factor of c.

Changing the sign of either the dependent or independent variable will reflect the graph about an axis.

Function	Effect on graph of $y = f(x)$
$y = -f(x)$	Reflect about x-axis.
$y = f(-x)$	Reflect about y-axis.

These operations may be combined to produce the graph of a "complex" function from the graph of a simpler function.

5) Given the graph below, sketch each:

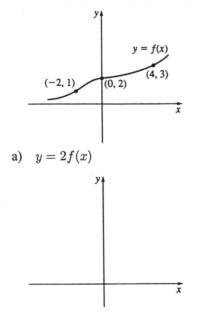

a) $y = 2f(x)$

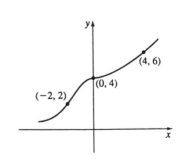

b) $y = f(x) - 2$

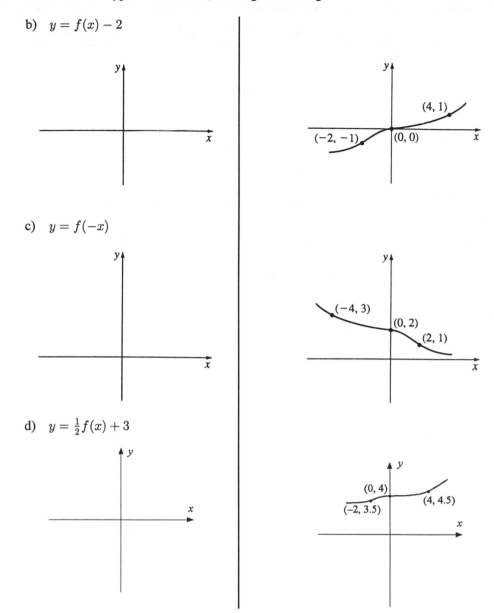

c) $y = f(-x)$

d) $y = \frac{1}{2}f(x) + 3$

Graphing Calculators and Computers

Concepts to Master

Determining an appropriate viewing rectangle.

Summary and Focus Questions

A <u>viewing rectangle</u> for a function $y = f(x)$ is the rectangle:

$$\{(x, y) \mid a \le x \le b,\ c \le y \le d\}$$

corresponding to a calculator or computer screen within which a portion of the graph of $y = f(x)$ is displayed. Selecting the correct window coordinates can be tricky -- choose a, b, c, and d so that the display brings out the most salient features of the graph.

Example: Choose an appropriate viewing window for $f(x) = 12 - x^2$.

Since the graph will have no y values greater than 12 ($(0, 12)$ is a high point -- a salient feature), $d = 13$ is a good choice. The graph is symmetric about the y-axis and crosses the x-axis (other salient features) between 3 and 4 and between -4 and -3. Good choices for a and b are $a = -5$ and $b = 5$. $f(5) = -13$, so $c = -15$ will work. The graph and viewing window are shown below:

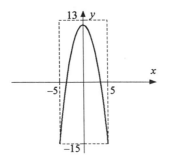

window: $[-5, 5]$ by $[-15, 13]$

12

1) Find an appropriate viewing rectangle for each:

   a) $y = f(x)$ whose graph is below:

   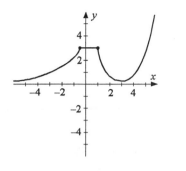

Your choices may vary. A good viewing window is $-4 \le x \le 5$, $-1 \le y \le 5$, or $[-4, 5]$ by $[-1, 5]$.

2) Is $[-50, 50]$ by $[-100, 100]$ an appropriate viewing rectangle for $y = \sin x$?

No. Since $-1 \le y \le 1$, the y scale is too large. Also, since sin is periodic, only a few periods need to be displayed. A window such as $[-12, 12]$ by $[-2, 2]$ would be more appropriate.

Principles of Problem Solving

Concepts to Master

Develop a systematic approach to problem solving.

Summary and Focus Questions

There is no single approach that will always work for solving problems in mathematics. However, this general strategy will be helpful:

1. Understand the problem -- determine what is known and unknown, draw a diagram, and introduce notation.
2. Make a plan -- look for the familiar, patterns, analogs, cases, and subcases. Look at the problem from two different perspectives.
3. Carry out the plan.
4. Look back over what you have done for errors.

Here are a couple of samples on which to practice this strategy:

1) What is the distance from the right angle to the hypotenuse in a right triangle with legs 5 and 12 inches?

1. We will call the unknown distance h and draw this figure to help:

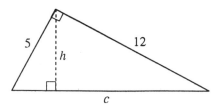

2. Our plan will be to calculate the area of the triangle two different ways and equate the expressions to get an equation involving h.

3. The area is $\frac{1}{2}5(12) = 30$. The hypotenuse is $c = \sqrt{5^2 + 12^2} = 13$. Thus a second way to calculate the area is $\frac{1}{2}h(13) = 6.5h$. Thus $6.5h = 30$ and $h = 4.615$.

4. The answer seems reasonable because it is a little less than 5 -- the hypotenuse in the smaller right triangle in the figure.

2. Al starts jogging at 10:00 at 5 mph. At 10:30 Bill starts jogging after Al at 6.5 mph. How long will it take for Bill to catch up to Al?

1. Let t be the time (in minutes) that Bill needs to catch up. Then $t + 30$ is the amount of time Al has spent jogging when Bill catches him.

2. Al and Bill will have travelled the same distance when Bill catches up. We calculate each distance in terms of t, equate them, and solve for t.

3. Bill's distance is $6.5t$.
    Al's distance is $5(t + 30)$.
    $6.5t = 5(t + 30)$
    $6.5t = 5t + 150$
    $1.5t = 150$
    $t = 100$ min. $= 1$ hr. 40 min.

4. An hour and 40 minutes seems like a reasonable answer.

1

Limits and Rates of Change

Cartoons courtesy of Sidney Harris. Used by permission.

The Tangent and Velocity Problems

Concepts to Master

A. Slope of secant line; Slope of tangent line
B. Interpretations of slopes as velocities

Summary and Focus Questions

A. A <u>secant line</u> at $P(a, b)$ on the graph of $f(x)$ is a line joining P and some other point Q also on $y = f(x)$. To find the slope of the tangent line at P, we select points Q closer and closer to P. Then the slope of the secant line becomes a better and better estimate of the slope of the tangent line at P. Finally our guess of what value the slopes of the secant lines approach is what we take as the slope of the tangent line.

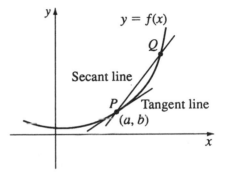

Once we know the slope m of the tangent line, the equation (using P as a point on the line) is $y - b = m(x - a)$.

1) Let P be the point $(3, 10)$ and Q be differing points on the graph of $y = f(x)$ given in the table. Find the slope of the secant lines PQ.

x	6	5	4	3.5	3.1
y	18	15	12.3	11.1	10.21

17

Q	Slope of PQ
$(6,\ 18)$	$\frac{18-10}{6-3} = 2.66$
$(5,\ 15)$	$\frac{15-10}{5-3} = 2.5$
$(4,\ 12.3)$	$\frac{12.3-10}{4-3} = 2.3$
$(3.5,\ 11.1)$	$\frac{11.1-10}{3.5-3} = 2.2$
$(3.1,\ 10.21)$	$\frac{10.21-10}{3.1-3} = 2.1$

2) Let $P(1,\ 5)$ be a point on the graph of $f(x) = 6x - x^2$. Let $Q(x,\ 6x - x^2)$ be on the graph. Find the slope of the secant line PQ for each given x value for Q.

x	$f(x)$	Slope of PQ
3		
2		
1.5		
1.01		

x	$f(x)$	Slope of PQ
3	9	$\frac{9-5}{3+1} = 2$
2	8	$\frac{8-5}{2-1} = 3$
1.5	6.75	$\frac{6.75-5}{1.5-1} = 3.5$
1.01	5.0399	$\frac{5.0399-5}{1.01-1} = 3.99$

3) Use your answer to question 2 to guess the slope of the tangent to $f(x)$ at P.

The values appear to be approaching 4.

4) What is the equation of the tangent line at P in question 2?

$y - 5 = 4(x - 1)$.

B. If $f(x)$ is interpreted as the distance an object is located from the origin along an x-axis at time x, then:

a) the slope of the secant line from P to Q is the <u>average velocity</u> from P to Q.

b) the slope of the tangent line at P is the <u>instantaneous velocity</u> at P.

Thus, calculating average velocity is performed in the same manner as calculating the slope of a secant line and calculating instantaneous velocity is the same as calculating the slope of a tangent line.

5) If a ball is $x^2 + 3x$ feet from the origin at any time x (in seconds), what is the instantaneous velocity when $x = 2$?

At $x = 2$, $f(2) = 2^2 + 3(2) = 10$.
Choose some x values approaching 2:

x	$f(x)$	Slope of Secant
3	18	$\frac{18-10}{3-2} = 8$
2.5	13.75	$\frac{13.75-10}{2.5-2} = 7.5$
2.1	10.71	$\frac{10.71-10}{2.1-2} = 7.1$

We guess the slope of the tangent, and thus the instantaneous velocity, is 7 ft/sec.

6) Let $f(x) = mx + b$, a linear position function. Sometimes, Always, or Never:
The average velocity for $f(x)$ from P to Q is the same as the instantaneous velocity at P.

Always.

The Limit of a Function

Concepts to Master

A. Limits; Numerical estimates; Graphical estimates
B. Right- and left-hand limits; Relationships between limits and one-sided limits
C. Infinite limits; Vertical asymptotes

Summary and Focus Questions

A. $\lim\limits_{x \to a} f(x) = L$ means that as x gets closer and closer to the number a (but not equal to a), the corresponding $f(x)$ values get closer and closer to the number L. (A more precise definition is in section 1.4.) For now, ignore the value $f(a)$, if there is one, in determining a limit.

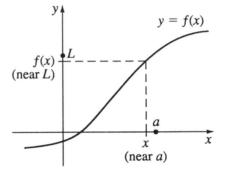

$\lim\limits_{x \to a} f(x) = L$ is rather like a guarantee: if you choose x close enough to a on the x-axis, then $f(x)$ is guaranteed to be close to L on the y-axis.

To say that "$\lim\limits_{x \to a} f(x)$ exists" means that there is some number L such that $\lim\limits_{x \to a} f(x) = L$.

To evaluate $\lim\limits_{x \to a} f(x)$ from the graph of $y = f(x)$, simply note what value corresponding $f(x)$ values approach on the y-axis as x values are chosen near a on the x-axis.

A somewhat risky method of calculating $\lim\limits_{x \to a} f(x)$ is to select several values of x near a and observe the pattern of corresponding $f(x)$ values. Then guess the value $f(x)$ is approaching. Your answer may vary depending on the particular values of x used and whether you experience rounding errors from your calculator or computer.

1) Answer each using the graph below:

a) $\lim\limits_{x \to -5} f(x) =$

5.

b) $\lim\limits_{x \to 2} f(x) =$

3. (It does not matter that $f(2) = 5$.)

c) $\lim\limits_{x \to 0} f(x) =$

Does not exist.

d) $\lim\limits_{x \to 4} f(x) =$

0.

2) Find $\lim\limits_{x \to 2} |x - 5|$ by making a table of values for x near 2.

x	$f(x) = |x - 5|$
3	2
2.1	2.9
1.99	3.01
2.0003	2.9997

We guess that $\lim\limits_{x \to 2} |x - 5| = 3$.

3) Estimate $\lim\limits_{x \to 1} \frac{1}{1-x}$.

We make a table of values:

x	$f(x) = \frac{1}{1-x}$
1.5	2
1.01	-100
0.99	100
1.0004	-2500

The values of $f(x)$ do not seem to congregate near a particular value. We guess that $\lim\limits_{x \to 1} \frac{1}{1-x}$ does not exist.

B. $\lim\limits_{x \to a^-} f(x) = L$, <u>the left-hand limit of f at a</u>, means that $f(x)$ gets closer and closer to L as x approaches a and $x < a$. It is called "left-hand" because it only concerns values of x less than a (to the left of a on the x-axis).

$\lim\limits_{x \to a^+} f(x) = L$, <u>the right-hand limit</u>, is a similar concept but involves only values of x greater than a.

If $\lim\limits_{x \to a^-} f(x)$ and $\lim\limits_{x \to a^+} f(x)$ both exist and are the same number L, then $\lim\limits_{x \to a} f(x)$ exists and is L.

If either one-sided limit does not exist or if they both exist but are different numbers, then $\lim\limits_{x \to a} f(x)$ does not exist. If $\lim\limits_{x \to a} f(x) = L$, then both one-sided limits exist and are equal L.

4) Answer each using the graph below:

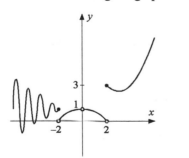

a) $\lim\limits_{x \to -2^-} f(x) = $ _____.

1.

b) $\lim\limits_{x \to -2^+} f(x) = $ _____.

0.

c) $\lim\limits_{x \to -2} f(x) = $ _____.

Does not exist.

d) $\lim\limits_{x \to 2^+} f(x) = $ _____.

3.

e) $\lim\limits_{x \to 2^-} f(x) = $ _____.

0.

f) $\lim\limits_{x \to 2} f(x) = $ _____.

Does not exist.

g) $\lim\limits_{x \to 0^-} f(x) = $ _____.

1.

h) $\lim\limits_{x \to 0^+} f(x) = $ _____.

1.

i) $\lim\limits_{x \to 0} f(x) = $ _____.

1.

5) $\lim\limits_{x \to 3^-} \frac{3-x}{|x-3|} = $ _____.

1. For x near 3 with $x < 3$, $x - 3 < 0$.
Thus $|x - 3| = -(x - 3) = 3 - x$.
Therefore $\lim\limits_{x \to 3^-} \frac{3-x}{|x-3|} = \lim\limits_{x \to 3^-} \frac{3-x}{3-x} = 1$.

6)   Find $\lim\limits_{x \to 2} f(x)$, where

$$f(x) = \begin{cases} 3x + 1 & \text{if } x < 2 \\ 8 & \text{if } x = 2 \\ x^2 + 3 & \text{if } x > 2 \end{cases}$$

$\lim\limits_{x \to 2^+} f(x) = \lim\limits_{x \to 2^+} x^2 + 3 = 7$.
$\lim\limits_{x \to 2^-} f(x) = \lim\limits_{x \to 2^-} 3x + 1 = 7$.
Since both one-sided limits are 7,
$\lim\limits_{x \to 2} f(x) = 7$.

7) Sometimes, Always, or Never:
    If $\lim\limits_{x \to 2} f(x)$ does not exist, then
    $\lim\limits_{x \to 2^+} f(x)$ does not exist.

Sometimes.
True in the case $f(x) = \frac{1}{x-2}$.
False in the case $f(x) = \frac{x-2}{|x-2|}$.

8) Sometimes, Always, or Never:
    If $\lim\limits_{x \to 2^+} f(x)$ does not exist, then
    $\lim\limits_{x \to 2} f(x)$ does not exist.

Always.

C. Sometimes $\lim\limits_{x \to a} f(x)$ does not exist because as x is assigned values that approach a, the corresponding $f(x)$ values grow larger without bound. In this case we say $\lim\limits_{x \to a} f(x) = \infty$. This still means $\lim\limits_{x \to a} f(x)$ does not exist, but it fails to exist in this special way.

$\lim\limits_{x \to a} f(x) = -\infty$ means $f(x)$ becomes smaller without bound as x approaches a.

As with ordinary limits, for $\lim\limits_{x \to a^+} f(x) = \pm\infty$, consider only $x > a$, and for $\lim\limits_{x \to a^-} f(x) = \pm\infty$, consider only $x < a$.

A vertical asymptote for $y = f(x)$ is a vertical line $x = a$ for which at least one of these limits hold:

$$\lim\limits_{x \to a^-} f(x) = \infty, \quad \lim\limits_{x \to a^-} f(x) = -\infty, \quad \lim\limits_{x \to a^+} f(x) = \infty, \quad \lim\limits_{x \to a^+} f(x) = -\infty.$$

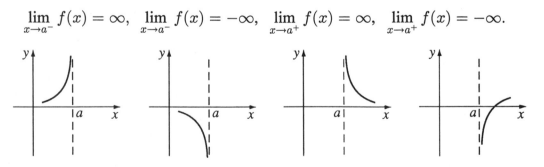

A function may have any number of vertical asymptotes or none at all. Some, like $y = \tan x$, have an infinite number.

9) Evaluate the limits given the graph of $y = f(x)$ below:

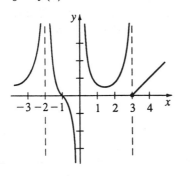

a) $\displaystyle\lim_{x \to -2^+} f(x) =$ _____.

∞.

b) $\displaystyle\lim_{x \to -2^-} f(x) =$ _____.

∞.

c) $\displaystyle\lim_{x \to -2} f(x) =$ _____.

∞.

d) $\displaystyle\lim_{x \to 3^-} f(x) =$ _____.

∞.

e) $\displaystyle\lim_{x \to 3^+} f(x) =$ _____.

0.

f) $\displaystyle\lim_{x \to 3} f(x) =$ _____.

Does not exist.

g) $\displaystyle\lim_{x \to 0^+} f(x) =$ _____.

∞.

h) $\displaystyle\lim_{x \to 0^-} f(x) =$ _____.

∞.

i) $\displaystyle\lim_{x \to 0} f(x) =$ _____.

Does not exist.

10) $\lim\limits_{x \to 3} \frac{1}{|x-3|} = $ _____.

∞. For x near 3, but not equal to 3, $|x - 3|$ is a small positive number. Therefore, $\frac{1}{|x-3|}$ is a large positive number.

11) True or False:
If $\lim\limits_{x \to a} f(x) = \infty$, and $\lim\limits_{x \to a} g(x) = \infty$,
then $\lim\limits_{x \to a} (f(x) - g(x)) = 0$.

False. For example, $f(x) = \frac{3}{x^2}$ and $g(x) = \frac{2}{x^2}$. Then $\lim\limits_{x \to 0} \frac{3}{x^2} = \infty$,
$\lim\limits_{x \to 0} \frac{2}{x^2} = \infty$, but $\lim\limits_{x \to 0} \left(\frac{3}{x^2} - \frac{2}{x^2} \right) = \lim\limits_{x \to 0} \frac{1}{x^2} = \infty$ (not 0!).

12) $\lim\limits_{x \to 3} \frac{x}{3-x}$

For x near 3, $\frac{x}{3-x}$ is the result of a number near 3 divided by a number near 0. Consider one-sided limits:
$\lim\limits_{x \to 3^+} \frac{x}{3-x} = -\infty$ because the denominator is negative, while $\lim\limits_{x \to 3^-} \frac{x}{3-x} = \infty$. Thus, $\lim\limits_{x \to 3} \frac{x}{3-x}$ does not exist.

13) What are the vertical asymptotes for the function in question 9?

$x = -2$, $x = 0$, and $x = 3$.

14) Find the vertical asymptotes of $f(x) = \frac{x}{|x|-1}$.

Good places to check are where $f(x)$ is not defined : at 1 and -1.
$\lim\limits_{x \to 1^+} \frac{x}{|x|-1} = \infty$, and $\lim\limits_{x \to 1^-} \frac{x}{|x|-1} = \infty$, so $x = 1$ and $x = -1$ are vertical asymptotes.

Section 1.3

Calculating Limits Using the Limit Laws

Concepts to Master

A. Evaluation of limits of rational functions at points in their domain
B. Limit laws (theorems) for evaluating limits
C. Algebraic manipulations prior to using the limit laws
D. Squeeze Theorem

Summary and Focus Questions

A. Limits of some functions at some points are easy to find:

If $f(x)$ is rational and a is in the domain of f, then $\lim\limits_{x \to a} f(x) = f(a)$.

For example, $\lim\limits_{x \to 3} \frac{x}{x+2} = \frac{3}{3+2} = \frac{3}{5}$ since 3 is in the domain of $f(x)$. Since polynomials are rational functions, the above applies to them as well, as in
$$\lim\limits_{x \to 2} (4x^3 - 6x^2 + 10x + 2) = 4(2)^3 - 6(2)^2 + 10(2) + 2 = 30.$$

1) Find $\lim\limits_{x \to 6} (4x^2 - 10x + 3)$.

$4(6)^2 - 10(6) + 3 = 87.$

2) Find $\lim\limits_{x \to 1} \frac{x^2 - 6x + 5}{x - 1}$.

Since 1 is not in the domain, we cannot substitute $x = 1$ to find this limit. The substitution yields $\frac{0}{0}$ which is a signal that the limit might still exist but must be found by a different method (see question 10). This limit turns out to be -4.

3) Find $\lim\limits_{x\to 2} \frac{x^2-6x+5}{x-1}$.

Unlike question 2, this one is straight forward. $\lim\limits_{x\to 2} \frac{x^2-6x+5}{x-1} = \frac{2^2-6(2)+5}{2-1} = -3$.

B. The eleven basic limit laws of this section, when applicable, allow you to evaluate complex limits by rewriting them as algebraic combinations of simpler limits of rational functions, as in:

$$\lim_{x\to 3} \sqrt{x^2+x} = \sqrt{\lim_{x\to 3}(x^2+x)} = \sqrt{\lim_{x\to 3}x^2 + \lim_{x\to 3}x} = \sqrt{9+3} = \sqrt{12}$$

The above was carried out to extremes to demonstrate the limit laws; since x^2+x is a polynomial, we could have concluded directly that $\lim\limits_{x\to 3}(x^2+x) = 12$.

The limit laws fail when a zero denominator results (see part C for what to do in such cases).

4) True or False:
    $\lim\limits_{x\to a}(f(x)g(x)) = \lim\limits_{x\to a}f(x)\lim\limits_{x\to a}g(x)$.

True, provided the limits exist.

5) True or False:
    $\lim\limits_{x\to a}f(g(x)) = \lim\limits_{x\to a}f(x)\lim\limits_{x\to a}g(x)$.

False.

6) Evaluate $\lim\limits_{x\to 5} \sqrt[3]{\frac{4x+44}{6x-29}}$.

$\lim\limits_{x\to 5} \sqrt[3]{\frac{4x+44}{6x-29}} = \sqrt[3]{\lim\limits_{x\to 5}\frac{4x+44}{6x-29}} = \sqrt[3]{\frac{64}{1}} = 4$.

7) Evaluate $\lim\limits_{x\to 1} \frac{x^2(6x+3)(2x-7)}{(x^3+4)(x+17)}$.

Do not multiply the numerator and denominator out. Applying the limit laws, this limit is
$$\frac{\left(\lim\limits_{x\to 1}x^2\right)\left(\lim\limits_{x\to 1}(6x+3)\right)\left(\lim\limits_{x\to 1}(2x-7)\right)}{\left(\lim\limits_{x\to 1}(x^3+4)\right)\left(\lim\limits_{x\to 1}(x+17)\right)}$$

$$= \frac{(1)(9)(-5)}{(5)(18)} = -\frac{1}{2}.$$

8) Find $\lim\limits_{x \to 3} f(x)$ where

$$f(x) = \begin{cases} x^2 & \text{if } x \geq 3 \\ 6x - 4 & \text{if } x < 3 \end{cases}$$

Because $f(x)$ is defined piecewise, we need to consider one-sided limits.
$$\lim\limits_{x \to 3^+} f(x) = \lim\limits_{x \to 3} x^2 = 9.$$
$$\lim\limits_{x \to 3^-} f(x) = \lim\limits_{x \to 3} 6x - 4 = 8.$$
Since the one-sided limits do not agree, $\lim\limits_{x \to 3} f(x)$ does not exist.

C. To find a limit, first try the limit laws. If they fail, as in
$$\lim\limits_{x \to 4} \frac{x^2 - 16}{x - 4} \quad \text{(gives } \tfrac{0}{0}, \text{ no information)},$$
you should next resort to algebraic techniques to rewrite the function in a form for which the limit theorems do work.

Division by zero is a common reason why the limit laws fail. Try to rewrite the function in a form that will not give a zero denominator. There are several techniques for this and often more than one technique is required in a given problem. The most common techniques are:

Cancellation:
When substitution of $x = a$ in $\lim\limits_{x \to a} \frac{f(x)}{g(x)}$ results in $\tfrac{0}{0}$, $x - a$ may be a factor of both $f(x)$ and $g(x)$ and hence can be canceled.
Find $\lim\limits_{x \to 2} \frac{2x^2 - 4x}{x^2 - 4}$.
Here $x = 2$ gives $\tfrac{0}{0}$ but $\frac{2x^2 - 4x}{x^2 - 4} = \frac{2x(x-2)}{(x+2)(x-2)} = \frac{2x}{x+2}$, provided $x \neq 2$.
Thus, $\lim\limits_{x \to 2} \frac{2x^2 - 4x}{x^2 - 4} = \lim\limits_{x \to 2} \frac{2x}{x+2} = \frac{2 \cdot 2}{2+2} = 1.$

Fraction Manipulation:
Some limits involve complex fractions. Rewriting the fraction in simplier terms can then allow the limit laws to be used.
Find $\lim\limits_{x \to 1} \frac{\frac{1}{x} - x}{\frac{1}{x} - 1}$.
We rewrite the function by first finding a simpler numerator and denominator and then using cancellation:

$$\frac{\frac{1}{x}-x}{\frac{1}{x}-1}=\frac{\frac{1}{x}-\frac{x^2}{x}}{\frac{1}{x}-1}=\frac{\frac{1-x^2}{x}}{\frac{1-x}{x}}=\frac{1-x^2}{1-x}=\frac{(1-x)(1+x)}{1-x}=1+x, \text{ for } x \neq 1.$$

Thus, $\displaystyle\lim_{x\to 1}\frac{\frac{1}{x}-x}{\frac{1}{x}-1}=\lim_{x\to 1}(1+x)=2.$

Rationalizing an Expression:

Sometimes if two radicals appear in the function, a process similar to using the conjugate of a complex number may be used to rewrite part of the function without radicals. For functions with $\sqrt{a}-\sqrt{b}$, use $\frac{\sqrt{a}+\sqrt{b}}{\sqrt{a}+\sqrt{b}}$. The result may be simplified using cancellation.

Find $\displaystyle\lim_{x\to 0}\frac{\sqrt{2+x}-\sqrt{2}}{x}.$

Here $\frac{0}{0}$ results, but if $\sqrt{2+x}-\sqrt{2}$ is multiplied by $\sqrt{2+x}+\sqrt{2}$, no radicals will remain in the numerator:

$$\frac{\sqrt{2+x}-\sqrt{2}}{x}\cdot\frac{\sqrt{2+x}+\sqrt{2}}{\sqrt{2+x}+\sqrt{2}}=\frac{2+x-2}{x(\sqrt{2+x}+\sqrt{2})}=\frac{1}{\sqrt{2+x}+\sqrt{2}}.$$ We end up with radicals in the denominator, but now we can use the limit laws.

$$\lim_{x\to 0}\frac{\sqrt{2+x}-\sqrt{2}}{x}=\lim_{x\to 0}\frac{1}{\sqrt{2+x}+\sqrt{2}}=\frac{1}{\sqrt{2}+\sqrt{2}}=\frac{1}{2\sqrt{2}}.$$

9) Evaluate $\displaystyle\lim_{x\to 6}\frac{2x-12}{x^2-x-30}.$

$$\frac{2x-12}{x^2-x-30}=\frac{2(x-6)}{(x-6)(x+5)}=\frac{2}{x+5}, \text{ for } x \neq 6.$$

$$\lim_{x\to 6}\frac{2}{x+5}=\frac{2}{11}.$$

10) Evaluate $\displaystyle\lim_{x\to 1}\frac{x^2-6x+5}{x-1}.$

$$\frac{x^2-6x+5}{x-1}=\frac{(x-1)(x-5)}{x-1}=x-5, \text{ for } x \neq 1.$$

$$\lim_{x\to 1}x-5=-4.$$

11) Evaluate $\displaystyle\lim_{x\to 3}\left[\frac{2x^2}{x-3}+\frac{6x}{3-x}\right].$

$$\frac{2x^2}{x-3}+\frac{6x}{3-x}=\frac{2x^2}{x-3}-\frac{6x}{x-3}=\frac{2x^2-6x}{x-3}$$

$$=\frac{2x(x-3)}{x-3}=2x, \text{ for } x \neq 3.$$

$$\lim_{x\to 3}2x=6.$$

12) Evaluate $\displaystyle\lim_{x\to 0}\frac{\frac{1}{4+x}-\frac{1}{4}}{x}.$

$$\frac{\frac{1}{4+x}-\frac{1}{4}}{x}=\frac{\frac{4}{(4+x)4}-\frac{(4+x)}{(4+x)4}}{x}=\frac{\frac{4-(4+x)}{(4+x)4}}{x}$$

$$=\frac{\frac{-x}{(4+x)4}}{x}=\frac{-1}{(4+x)4}.$$

$$\lim_{x\to 0}\frac{-1}{(4+x)4}=-\frac{1}{16}.$$

13) Evaluate $\lim\limits_{x \to 3} \frac{\sqrt{x}-\sqrt{3}}{x-3}$.

$$\frac{\sqrt{x}-\sqrt{3}}{x-3} \cdot \frac{\sqrt{x}+\sqrt{3}}{\sqrt{x}+\sqrt{3}} = \frac{x-3}{(x-3)\left(\sqrt{x}+\sqrt{3}\right)}$$
$$= \frac{1}{\sqrt{x}+\sqrt{3}}, \text{ for } x \neq 3.$$
$$\lim\limits_{x \to 3} \frac{1}{\sqrt{x}+\sqrt{3}} = \frac{1}{2\sqrt{3}}.$$

D. If $g(x)$ is "trapped" between $f(x)$ and $h(x)$ (i.e., if $f(x) \le g(x) \le h(x)$ for all x near a, except $x = a$) then
$$\lim\limits_{x \to a} f(x) \le \lim\limits_{x \to a} g(x) \le \lim\limits_{x \to a} h(x), \text{ if each limit exists.}$$
In addition, if $\lim\limits_{x \to a} f(x) = \lim\limits_{x \to a} h(x)$ then $\lim\limits_{x \to a} g(x)$ must exist and be that same number (the Squeeze Theorem).

14) True or False:
   If $f(x) < g(x)$ for all x then
   $$\lim\limits_{x \to a} f(x) < \lim\limits_{x \to a} g(x).$$

False. Let $a = 0$, $f(x) = -x^2$ and
$$g(x) = \begin{cases} x^2 & \text{if } x \neq 0 \\ 1 & \text{if } x = 0 \end{cases}.$$
$f(x) < g(x)$, for all x but
$$\lim\limits_{x \to 0} f(x) = \lim\limits_{x \to 0} -x^2 = 0 = \lim\limits_{x \to 0} x^2$$
$$= \lim\limits_{x \to 0} g(x).$$

15) For $0 \le x \le 1$, $x + 1 \le 3^x \le 2x + 1$.
   Use this to find $\lim\limits_{x \to 0^+} 3^x$.

Since $\lim\limits_{x \to 0^+} (x + 1) = 1$ and
$\lim\limits_{x \to 0^+} (2x + 1) = 1$, by the Squeeze
Theorem, $\lim\limits_{x \to 0^+} 3^x = 1$.

The Precise Definition of a Limit

Concepts to Master

A. Epsilon-delta (ϵ-δ) definition of $\lim\limits_{x \to a} f(x) = L$

B. Definitions of infinite limits

Summary and Focus Questions

A. $\lim\limits_{x \to a} f(x) = L$ means for any given $\epsilon > 0$, there exists $\delta > 0$ such that
    if $0 < |x - a| < \delta$, then $|f(x) - L| < \epsilon$.

This is consistent with the intuitive definition given earlier if you think of ϵ and δ as small numbers and note that $|f(x) - L|$ is the distance between $f(x)$ and L while $|x - a|$ is the distance between x and a. This definition says that $f(x)$ will always be within ϵ units of $L(|f(x) - L| < \epsilon)$ whenever x is within δ units of $a(|x - a| < \delta)$.

A proof that $\lim\limits_{x \to a} f(x) = L$ consists of a verification of the definition. Start by assuming you have a $\delta > 0$ and then determine a number δ so that the statement $|f(x) - L| < \epsilon$ can be deduced from the statement $0 < |x - a| < \delta$. There are no universal techniques for determining δ in terms of ϵ but often this is accomplished by starting with $|f(x) - L| < \epsilon$ and replacing it with equivalent statements until finally the statement $|x - a| < \delta$ results. The reversal of these steps constitutes a major portion of the proof.

Reducing $|f(x) - L| < \epsilon$ down to $|x - a| < \delta$ may require several rewritings of $|f(x) - L| < \epsilon$ using factoring and cancellation. You may also need properties of absolute value, the most common being
    $|a \cdot b| = |a| \, |b|$ and
    $|a| < b$ is equivalent to $-b < a < b$.

Example:
To prove $\lim\limits_{x \to 3} (6x - 5) = 13$ we start with $|f(x) - L|$ and simplify:
$$|f(x) - L| = |6x - 5 - 13| = |6x - 18| = |6(x - 3)|$$
$$= |6|\,|x - 3| = 6|x - 3|.$$

Thus, $|6x - 5 - 13| < \epsilon$ is equivalent to $6|x - 3| < \epsilon$.
Divide by 6: $|x - 3| < \frac{\epsilon}{6}$. Our choice for δ is $\frac{\epsilon}{6}$.
This scratch work gives us the details of the proof.

Proof:
Let $\epsilon > 0$. Choose $\delta = \frac{\epsilon}{6}$.
Then for $0 < |x - 3| < \delta$
$$|x - 3| < \tfrac{\epsilon}{6}$$
$$6|x - 3| < \epsilon$$
$$|6x - 18| < \epsilon$$
$$|6x - 5 - 13| < \epsilon$$
In summary, if $0 < |x - 3| < \delta$, then $|6x - 5 - 13| < \epsilon$.
Thus, $\lim\limits_{x \to 3} (6x - 5) = 13$.

When $f(x)$ is nonlinear, a simplified $|f(x) - L| < \epsilon$ may be in the form
$|M(x)|\,|x - a| < \epsilon$ where $M(x)$ is some expression involving x. The
procedure here is to find a constant K such that $|M(x)| < K$ for all x such
that $|x - a| < p$ for some positive number p (often $p = 1$). The choice of δ is
then the smaller of p and $\frac{\epsilon}{K}$, written $\delta = \min\left\{p, \frac{\epsilon}{K}\right\}$.

Example:
To prove $\lim\limits_{x \to 4} x^2 - 1 = 15$, we have
$$|f(x) - L| = |x^2 - 1 - 15| = |x^2 - 16| = |(x + 4)(x - 4)| = |x + 4||x - 4|.$$
For $p = 1$, if $|x - 4| < p,$ then $\quad |x - 4| < 1$
$$-1 < x - 4 < 1$$
$$-3 < x < 5$$
$$1 < x + 4 < 9$$
Thus, $|x + 4| < 9$. Therefore, using $K = 9$, when $|x - 4| < 1$, we have
$|f(x) - L| = |x + 4||x - 4| < 9|x - 4|$ and $9|x - 4| < \epsilon$ if $|x - 4| < \frac{\epsilon}{9}$.
Thus choose $\delta = \min\left\{1, \frac{\epsilon}{9}\right\}$.

1) What part of the definition of
$\lim_{x \to a} f(x) = L$ ensures that $x = a$ is
not considered?

$$0 < |x - a|.$$

2) The choice of δ usually (does, does
not) depend upon the size of ϵ.

Does. For most limits, the smaller the ϵ
given, the smaller must be the
corresponding choice of δ.

3) For arbitrary $\epsilon > 0$, find a $\delta > 0$, in
terms of ϵ such that the definition of
limit is satisfied for $\lim_{x \to a} (2x + 1) = 9$.

$|f(x) - L| < \epsilon$ (Start)
$|2x + 1 - 9| < \epsilon$ (Replacement)
$|2x - 8| < \epsilon$ (Simplification)
$2|x - 4| < \epsilon$ (Factoring)
$|x - 4| < \frac{\epsilon}{2}$ (Divide by 2)
The desired δ is $\delta = \frac{\epsilon}{2}$.

4) Find δ for an arbitrary ϵ in
$\lim_{x \to 3} x^2 - 4 = 5$.

$|f(x) - L| = |x^2 - 4 - 5| = |x^2 - 9|$
$\qquad = |(x + 3)(x - 3)| = |x + 3||x - 3|$
If $p = 1$ and $|x - 3| < 1$,
then $-1 < x - 3 < 1$
$\qquad 2 < x < 4$
$\qquad 5 < x + 3 < 7$
So $|x + 3| < 7$.
Let $\delta = \min\{1, \frac{\epsilon}{7}\}$.

B. Recall that $\lim_{x \to a} f(x) = \infty$ means that as x is assigned values that approach a,
the corresponding $f(x)$ values grow larger without bound. The definition to
make this precise is

$\lim_{x \to a} f(x) = \infty$, means for all positive M there is $\delta > 0$, such that

$f(x) > M$ whenever $0 < |x - a| < \delta$.

Just like with ordinary limits, you may need to work "backwards" from
$f(x) > M$ to determine δ.

5) Prove that $\lim\limits_{x \to 2} \frac{1}{|4-2x|} = \infty$.

For $M > 0$, $f(x) > M$ is $\frac{1}{|4-2x|} > M$.

$\frac{1}{M} > |4 - 2x|$

$|4 - 2x| < \frac{1}{M}$

$|(-2)(x - 2)| < \frac{1}{M}$

$2|x - 2| < \frac{1}{M}$

$|x - 2| < \frac{1}{2M}$

Choose $\delta = \frac{1}{2M}$.

Proof:

Let $M > 0$; choose $\delta = \frac{1}{2M}$.

Then if $0 < |x - 2| < \delta$,

$|x - 2| < \frac{1}{2M}$

$2|x - 2| < \frac{1}{M}$

$|2x - 4| < \frac{1}{M}$

$|4 - 2x| < \frac{1}{M}$

$M < \frac{1}{|4-2x|}$

$\frac{1}{|4-2x|} < M$.

Thus, if $0 < |x - 2| < \delta$, then $\frac{1}{|4-2x|} > M$.

Therefore, $\lim\limits_{x \to 2} \frac{1}{|4-2x|} = \infty$.

Continuity

Concepts to Master

A. Continuity of a function at a point and on an interval
B. Intermediate Value Theorem

Summary and Focus Questions

A. A function f is <u>continuous at a</u> means both numbers $f(a)$ and $\lim\limits_{x \to a} f(x)$ exist and are equal, that is, $\lim\limits_{x \to a} f(x) = f(a)$.

If f is continuous at a, then the calculation of $\lim\limits_{x \to a} f(x)$ is easy; just use $x = a$ in $f(x)$. Continuity at a means the graph of f is unbroken as it passes through $(a, f(a))$.

f is <u>continuous on an interval</u> means f is continuous at each point in the interval. The graph of $f(x)$ must be unbroken at each point in the interval. Algebraic (and therefore rational and polynomial) functions are continuous at each point in their domains.

1)

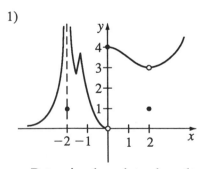

Determine the points where the function f above is not continuous (is discontinuous).

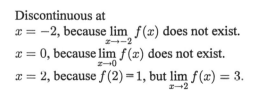

Discontinuous at
$x = -2$, because $\lim\limits_{x \to -2} f(x)$ does not exist.
$x = 0$, because $\lim\limits_{x \to 0} f(x)$ does not exist.
$x = 2$, because $f(2) = 1$, but $\lim\limits_{x \to 2} f(x) = 3$.

2) For the graph in problem 1 determine whether f is continuous on:

    a) $[-4, -3]$

        Yes.

    b) $[.5, 1.5]$

        Yes.

    c) $[-1, 1]$

        No, not continuous at 0.

    d) $[-2, -1]$

        Yes.

    e) $[0, 1]$

        Yes.

3) Where is $f(x) = \frac{x}{(x-1)(x+2)}$ continuous?

    Since f is rational, f is continuous everywhere it is defined: f is continuous for all x except $x = 1$, $x = -2$.

4) Is $f(x) = 4x^2 + 10x + 1$ continuous at $x = \sqrt{2}$?

    Yes. It is a polynomial, so f is continuous everywhere.

5) Where is $f(x) = \sqrt{\frac{1-x}{|x|}}$ continuous?

    $(-\infty, 0) \cup (0, 1]$, f is algebraic and this is the domain of f.

B. The Intermediate Value Theorem is: If f is continuous on $[a, b]$ and N is between $f(a)$ and $f(b)$ then there is at least one c between a and b so that $f(c) = N$.
Thus, as x varies from a to b, $f(x)$ must attain all values between $f(a)$ and $f(b)$.

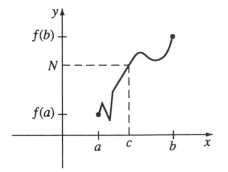

6) Does $f(x) = x^2 - x - 3$ have a root (a value w such that $f(w) = 0$) in $[1, 3]$?

Yes. Since f is a polynomial f is continuous on $[1, 3]$. Since 0 is between $-3 = f(1)$ and $3 = f(3)$, by the Intermediate Value Theorem, there exists w between 1 and 3 with $f(w) = 0$.

7) Is the hypothesis of the Intermediate Value Theorem satisfied for
$$f(x) = \begin{cases} \sin \frac{\pi}{x} & x \neq 0 \\ 0 & x = 0 \end{cases}$$
with domain $[-2, 2]$?

No. $f(x)$ is not continuous at 0, hence not continuous on $[-2, 2]$.

8) For $f(x) = x^2 - 3x$, is there a $t \in [-1, 5]$ such that $f(t) = -2$?

The answer is "yes", but we cannot use the Intermediate Value Theorem to support the answer. $f(-1) = 4$ and $f(5) = 10$, but -2 is not between 4 and 10. Solving $x^2 - 3x = -2$ gives $x = 1$, $x = 2$, both in $[-1, 5]$.

9) Find an interval $[a, b]$ for which $f(x) = x^5 - 10x^2 + bx - 1$ has a root.

Since f is a polynomial f is continuous everywhere. We simply need to find an a with $f(a) < 0$ and b with $f(b) > 0$. $a = 0$ and $b = 3$ will do. By the Intermediate Value Theorem f has a root in $[0, 3]$.

Tangents, Velocities, and Other Rates of Change

Concepts to Master

A. Slope of the tangent line to a graph
B. Instantaneous velocity; Rates of change

Summary and Focus Questions

A. If $m = \lim\limits_{x \to a} \frac{f(x) - f(a)}{x - a}$ exists, we say m is the slope of the tangent line to $y = f(x)$ at the point $(a, \, f(a))$.

If we use $a + h = x$, then as $x \to a$, we have $h \to 0$. Thus an alternate way to write the above limit is $m = \lim\limits_{h \to 0} \frac{f(a+h) - f(a)}{h}$.

Either form may be used to calculate the slope of the tangent line. Notice that the direct substitution $x = a$ or $h = 0$ in these limits always yields $\frac{0}{0}$, so you can never simply use limit laws to evaluate the slope of the tangent line.

1) Find the slope of the tangent line to $f(x) = x^2 + 2x$ at $x = 3$.

At $x = 3$, $f(3) = 3^2 + 2(3) = 15$.
$\frac{f(x) - f(a)}{x - a} = \frac{x^2 + 2x - 15}{x - 3} = \frac{(x-3)(x+5)}{x-3}$
$= x + 5$ for $x \neq 3$.
Thus, $\lim\limits_{x \to 3} \frac{f(x) - f(3)}{x - 3} = \lim\limits_{x \to 3}(x + 5) = 8$.

2) Find $\lim\limits_{h \to 0} \frac{f(a+h) - f(a)}{h}$ where $f(x) = \frac{1}{x-1}$ and $a = 2$.

$f(2 + h) = \frac{1}{2+h-1} = \frac{1}{h+1}$.
$f(2 + h) - f(2) = \frac{1}{h+1} - \frac{1}{2-1} = \frac{1}{h+1} - \frac{1}{1}$
$= \frac{1}{h+1} - \frac{h+1}{h+1} = \frac{1 - (h+1)}{h+1} = \frac{-h}{h+1}$.
Thus, $\lim\limits_{h \to 0} \frac{f(a+h) - f(a)}{h} = \lim\limits_{h \to 0} \frac{-h}{h+1} \cdot \frac{1}{h}$
$= \lim\limits_{h \to 0} \frac{-1}{h+1} = -1$.

B. The expression $\frac{f(x)-f(a)}{x-a}$ is the <u>average rate of change</u> of $y = f(x)$ on the interval $[x, a]$. Thus $\lim\limits_{x \to a} \frac{f(x)-f(a)}{x-a}$ is the <u>instantaneous rate of change</u> of $y = f(x)$ at $x = a$.

In particular, if $y = f(x)$ is the position of an object at time x, $\frac{f(x)-f(a)}{x-a}$ is the <u>average velocity</u> and $\lim\limits_{x \to a} \frac{f(x)-f(a)}{x-a}$ is the <u>instantaneous velocity</u> at $x = a$.

3)   Find the instantaneous velocity at time 3 seconds if a particle's position at time t is $f(t) = t^2 + 2t$ feet.

     8 ft/sec. This question is asking for the same information as question 1. Here the rate of change is interpreted as velocity instead of slope.

4)   a)   For a particle whose position at time t is $f(t) = 6t^2 - 4t + 1$ feet, find the average velocity over each of these intervals of time:

        $[1, 4]$

$$\frac{f(4)-f(1)}{4-1} = \frac{81-3}{3} = 26 \text{ ft/sec}$$

        $[1, 2]$

$$\frac{f(2)-f(1)}{2-1} = \frac{17-3}{1} = 14 \text{ ft/sec}$$

        $[1, 1.2]$

$$\frac{f(1.2)-f(1)}{1.2-1} = \frac{4.84-3}{0.2} = 9.2 \text{ ft/sec}$$

        $[1, 1.01]$

$$\frac{f(1.01)-f(1)}{1.01-1} = \frac{3.0806-3}{0.01} = 8.06 \text{ ft/sec}$$

    b)   Find the instantaneous velocity of the particle at $t = 1$ second.

     From part (a) it appears as though the answer is 8 ft/sec. Let's find out:

$$\frac{f(t)-f(1)}{t-1} = \frac{6t^2-4t+1-3}{t-1} = \frac{6t^2-4t-2}{t-1}$$
$$= \frac{(6t+2)(t-1)}{t-1} = 6t + 2 \text{ for } t \neq 1.$$

Thus,
$$\lim_{t \to 1} \frac{f(t)-f(1)}{t-1} = \lim_{t \to 1} (6t + 2) = 8 \text{ ft/sec.}$$

Derivatives

Cartoons courtesy of Sidney Harris. Used by permission.

Derivatives

Concepts to Master

A. Definition of the derivative; Notations
B. Interpretations of the derivative
C. Differentiability; Relationship of continuity to differentiability
D. Relationship between the graphs of f and f'

Summary and Focus Questions

A. The derivative of $y = f(x)$ at a point x is $f'(x) = \lim\limits_{h \to 0} \frac{f(x+h)-f(x)}{h}$.

All of these notations are used to refer to the derivate of f at x:
$$f'(x),\ y',\ \frac{df}{dx},\ \frac{dy}{dx},\ \frac{d}{dx}f(x),\ \mathbf{D}f(x),\ \mathbf{D}_x f(x)$$

For now you must calculate $f'(x)$ in the same manner as the limits of Section 1.6:

1) Find $f(x + h)$.
2) Find $f(x + h) - f(x)$.
3) Determine $\frac{f(x+h)-f(x)}{h}$ and simplify (if possible) when $h \neq 0$.
4) Find the limit of the result of part 3).

1) Find $f'(3)$ for $f(x) = x^2 + 10x$.

$f'(3)$ is found in steps:
(1) $f(3 + h) = (3 + h)^2 + 10(3 + h)$
$= 9 + 6h + h^2 + 30 + 10h$
$= 39 + 16h + h^2$.
(2) $f(3 + h) - f(3)$
$= (39 + 16h + h^2) - 39$
$= 16h + h^2 = h(16 + h)$.
(3) $\frac{f(3+h)-f(3)}{h} = \frac{h(16+h)}{h} = 16 + h$
for $h \neq 0$.
(4) Finally, $f'(3) = \lim\limits_{h \to 0}(16 + h) = 16$.

Handwritten work (left):
1) $f(3+h) = (3+h)^2 + 10(3+h)$
$= 9 + 6h + h^2 + 30 + 10h$
$= h^2 + 16h + 39$

2) $f(3th) = (h^2 + 16h + 39) - ((3)^2 + 30)$
$h^2 + 16h + 39 - 9 - 30$
$h^2 + 16h \quad 39 - 39$
$h^2 + 16h \quad h(h+16)$

42

Handwritten work (right):
3) $\frac{f(3+h) - f(3)}{h} = \frac{h(h+16)}{h} = 16 + h$
4) $\lim\limits_{h \to 0} 16 + h = 16$

2) Find $f'(x)$ for $f(x) = 2x - x^2$.

(1) $f(x+h) = 2(x+h) - (x+h)^2$
$= 2x + 2h - [[x^2 + 2hx + h^2]]$
$= 2x + 2h - x^2 - 2hx - h^2$

(2) $f(x+h) - f(x) = 2x + 2h - x^2 - 2hx + h^2$
$- (2x - x^2)$

$2x + 2h - x^2 - 2hx - h^2 - 2x + x^2$
$2h - 2hx - h^2$
$h(2 - 2x - h)$

(1) $f(x + h) = 2(x + h) - (x + h)^2$
$= 2x + 2h - x^2 - 2xh - h^2.$

(2) $f(x + h) - f(x)$
$= 2x + 2h - x^2 - 2xh - h^2$
$- (2x - x^2)$
$= 2h - 2xh - h^2$
$= h(2 - 2x - h).$

(3) $\frac{f(x+h)-f(x)}{h} = \frac{h(2-2x-h)}{h}$
$= 2 - 2x - h$ for $h \neq 0.$

(4) $f'(x) = \lim_{h \to 0} (2 - 2x - h)$
$= 2 - 2x - 0 = 2 - 2x.$

3) True or False:
For $y = f(x)$ the notation $f'(x)$ and $\frac{dx}{dy}$ have the same meaning.

False. $f'(x) = \frac{dy}{dx}.$

B. If $f'(a)$ exists, the tangent line to the graph of f at a is the line through $(a,\ f(a))$ with slope $f'(a)$. If $f'(a)$ does not exist, then the tangent line might not exist, might be a vertical line, or might not be unique.

The slope of the tangent line (measured by $f'(a)$) is the same as the instantaneous rate of change of $y = f(x)$ with respect to x at $x = a$.

4) The slope of the tangent line to the graph of f at the point a is $f'(a)$.

$f'(a)$, provided it exists.

5) True or False:
$f'(x)$ measures the average rate of change of $y = f(x)$ with respect to x.

False. $f'(x)$ measures instantaneous rate of change.

6) Given that $f'(x) = 3x^2 + 2$ for $f(x) = x^3 + 2x + 1$, find:

a) the equation of the tangent line to
    f at the point corresponding to
    $x = 2$.

> The point of tangency has x coordinate 2
> and y coordinate
> $f(2) = 2^3 + 2(2) + 1 = 13$.
> The slope is $f'(2) = 3(2)^2 + 2 = 14$.
> Hence the tangent line has equation
> $y - 13 = 14(x - 2)$ or $y = 14x - 15$.

b) the instantaneous rate of change of
    $f(x)$ with respect to $x = 3$.

> $f'(3) = 3(3)^2 + 2 = 29$.

c) If $f(x)$ represents the distance in
    feet of a particle from the origin at
    time x, find the velocity at time
    $x = 4$ seconds.

> $f'(4) = 3(4)^2 + 3 = 51$ ft/sec.

C. If f is differentiable at a point c, then f is also continuous at c. (Roughly
    speaking, this means that if you can draw a tangent line to a graph, then the
    graph must be unbroken.) The converse is *false*: continuity *does not* imply
    differentiability.

There are three common ways for a function to fail to be differentiable at a
point.

1) The graph has a corner or
    "kink" in it. Example:
    $$f(x) = \begin{cases} x^2 & , x \le 2 \\ (x-4)^2 & , x > 2 \end{cases}$$

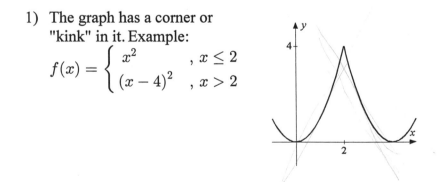

2) The graph is not continuous.
   Example:
   $$f(x) = \begin{cases} x^2 & , x \le 2 \\ 5 & , x = 2 \\ 10 - x^2 & , x > 2 \end{cases}$$

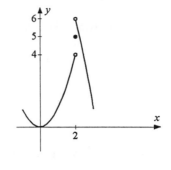

3) The graph has a tangent line,
   but it is a vertical line.
   Example: $f(x) = \sqrt[3]{x-2}$

7) Can a function be continuous at $x = 3$
   but not differentiable at $x = 3$?

   Yes. $f(x) = |x - 3|$ is an example.
   $f'(3)$ does not exist.

8) Can a function be differentiable at
   $x = 3$ and not continuous at $x = 3$?

   No. This can never happen.

9) True or False:
   For the function f graphed below:

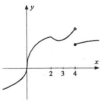

a) $f(x)$ is continuous at 0.

True.

b) $f(x)$ is differentiable at 0.

False.

c) $f(x)$ is continuous at 2.

True.

d) $f(x)$ is differentiable at 2.

False.

e) $f(x)$ is continuous at 3.

True.

f) $f(x)$ is differentiable at 3.

True.

g) $f(x)$ is continuous at 4.

False.

h) $f(x)$ is differentiable at 4.

False.

D. The graph of f' may be determined from the graph of f by remembering that $f'(x)$ is the slope of the tangent line at $(x, f(x))$. In the first graph below of $y = f(x)$, we have labeled some tangent line slope values. Beneath that graph is the corresponding graph of $y = f'(x)$.

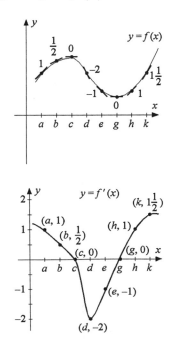

10) From the graph of f below estimate $f'(1)$, $f'(2)$, $f'(3)$, and $f'(4)$ and then sketch a graph of $y = f'(x)$.

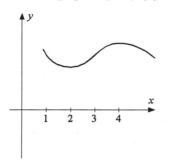

We estimate that $f'(1) = -2$, $f'(2) = 0$, $f'(3) = 1$, and $f'(4) = 0$. The graph of f' is rather like this:

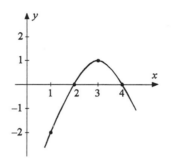

11) From the given graph of $y = f'(x)$ sketch a graph of $y = f(x)$.

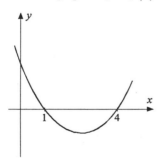

One such graph of $y = f(x)$ is:

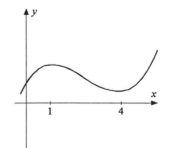

Differentiation Formulas

Concepts to Master

Differentiation from formulas rather than from limits

Summary and Focus Questions

The following differentiation rules must be memorized:

$\frac{dc}{dx} = 0$ for any constant c.

$\frac{d}{dx}(x^n) = nx^{n-1}$ for any real number. (General Power Rule.)

$[cf(x)]' = cf'(x)$ for any constant c.

$[f(x) \pm g(x)]' = f'(x) \pm g'(x)$.

$(f(x) \cdot g(x))' = f(x)g'(x) + g(x)f'(x)$.

$\left[\frac{f(x)}{g(x)}\right]' = \frac{g(x)f'(x) - f(x)g'(x)}{(g(x))^2}$

1) Find $f'(x)$ for each:

a) $f(x) = 10$

$f'(x) = 0.$

b) $f(x) = x^6$

$f'(x) = 6x^5.$

c) $f(x) = 5x^4$

$f'(x) = 20x^3.$

d) $f(x) = x^4 + 4x^3$

$f'(x) = 4x^3 + 12x^2.$

e) $f(x) = 7x^3 + 6x^2 + 10x + 12$

$f'(x) = 21x^2 + 12x + 10.$

f) $f(x) = (x^2 + x)(3x + 1)$

$f'(x) = (x^2 + x)(3) + (3x + 1)(2x + 1)$
$\quad = 9x^2 + 8x + 1$

$(x^2+x)(3) + (3x+1)(2x+1)$

$3x^2 + 3x + 6x^2 + 3x + 2x + 1$

$9x^2 + 8x + 1$

48

g) $f(x) = \frac{x^3}{x^2+10}$ $\frac{(x^2+10)(3x^2)-(x^3)(2x)}{(x^2+10)^2}$

$$f'(x) = \frac{(x^2+10)(3x^2)-(x^3)(2x)}{(x^2+10)^2} = \frac{x^4+30x^2}{(x^2+10)^2}$$

h) $f(x) = x^{-7} - x^{-6}$

$-7x^{-8} + 6x^{-7}$

$$f'(x) = -7x^{-8} + 6x^{-7}$$

i) $f(x) = x^\pi$ $\pi x^{\pi-1}$

$$f'(x) = \pi x^{\pi-1}$$

j) $f(x) = \frac{1+x^{-1}}{2-x^{-2}}$

$\frac{(2-x^{-2})(-x^{-2}) - (1+x^{-1})(2x^{-3})}{(2-x^{-2})^2}$

$$f'(x) = \frac{(2-x^{-2})(-x^{-2})-(1+x^{-1})(2x^{-3})}{(2-x^{-2})^2}$$

2) Suppose the position of a particle along an axis at time t is $f(t) = t^2 + \frac{1}{t}$ ft. Find the velocity of the particle at time $t = 2$ seconds.

$f'(t) = 2t + (-t)^{-2}$

$f'(t) = 2t - t^{-2}$

$= 2t - \frac{1}{t^2}$

$2(2) - \frac{1}{(2)^2}$

$4 - \frac{1}{4} = 3.75$

We can now use differentiation formulas to find velocities.
$f(t) = t^2 + t^{-1}$
$f'(t) = 2t + (-1)t^{-2} = 2t - \frac{1}{t^2}$
$f'(2) = 2(2) - \frac{1}{2^2} = 3.75$ ft/sec.

3) For $f(x) = mx + b$, find $f'(x)$. Interpret your result.

$f'(x) = m$ a constant

tangent lineslope is m always

$f'(x) = m$, a constant.
$f(x) = mx + b$ is a line with slope m.
Thus, the tangent line is the line itself and the tangent line slope is always m.

Rates of Change in the Natural and Social Sciences

Concepts to Master

Interpretation of a derivative as an instantaneous rate of change of a given quantity.

Summary and Focus Questions

For $y = f(x)$, $f'(x)$ measures the instantaneous rate of change of y with respect to x. Rates of change play an important role in many fields. If $f(x)$ represents some quantity at time x, $f'(x)$ will measure its instantaneous rate of change at the given time x.

1) Water is flowing out of a water tower in such a fashion that after t minutes there are $10000 - 10t - t^3$ gallons left. How fast is the water flowing after 2 minutes?

$V(t) = 10000 - 10t - t^3$

$V'(t) = -10 - 3t^2$

$= -10 - 3(2)^2$

$= -10 - 12 = -22 \text{ gal/min}$

Let $V(t) = 10000 - 10t - t^3$ be the volume at time t. The question asks for the rate of change of V.
$V'(t) = -10 - 3t^2$. At $t = 2$,
$V'(2) = -10 - 3(2)^2 = -22$ gal/min.
Note: The answer is negative, indicating that the volume of water is decreasing.

2) A space shuttle is $16t + t^3$ meters from its launch pad t seconds after liftoff. What is its velocity after 3 seconds?

$V(t) = 16t + t^3$

$V'(t) = 16 + 3t^2$

$= 16 + 3(3)^2$

$= 16 + 3(9)$

$= 16 + 27$

$= 43 \text{ m/s}$

$d(t) = 16t + t^3$, $d'(t) = 16 + 3t^2$
$d'(t) = 16 + 3(3)^2 = 43$ m/sec.

50

3) A particle is moving along an axis so that at time t its position is
$f(t) = t^3 - 6t^2 + 6$ feet.

a) What is its velocity at time t?

$f'(t) = 3t^2 - 12t$

$$f'(t) = 3t^2 - 12t$$

b) What is the velocity at 3 seconds?

$f'(3) = 3(3)^2 - 12(3)$
$= 27 - 36 = -9 \text{ ft/s}$

$$f'(3) = 3(3)^2 - 12(3) = -9 \text{ ft/sec.}$$

c) Is the particle moving left or right at 3 seconds?

moving left $\boxed{-9}$

Left, since $f'(3) = -9$ is negative.

d) At what time(s) is the particle (instantaneously) motionless?

$f'(t) = 0$

$3t^2 - 12t = 0$ $6, 4$

$3t(t - 4) = 0$

$0 \quad t = 4$

Motionless means its velocity
$f'(t) = 0$
$3t^2 - 12t = 0$
$3t(t - 4) = 0$
At $t = 0$ and $t = 4$ the particle has zero velocity.

4) A stone is thrown upward from a 70 m cliff so that its height above ground is $f(t) = 70 + 3t - t^2$. What is the velocity of the stone as it hits the ground?

$f(t) = 70 + 3t - t^2$

$0 = 70 + 3t - t^2$

$= (10 - t)(7 + t) = 0$

$\_ \ t = 10 \text{ or } \not{\!\!7}$

$f'(t) = 3 - 2t$

$= 3 - 2(10)$

$= 3 - 20$

$= -17 \text{ m/s}$

The time when the stone hits the ground is when $f(t) = 0$.
$70 + 3t - t^2 = 0$
$(10 - t)(7 + t) = 0$
$t = 10$ or $t = -7$. (Disregard $t = -7$.)
$f'(t) = 3 - 2t$.
$f'(10) = 3 - 20 = -17 \text{ m/sec.}$

Derivatives of Trigonometric Functions

Concepts to Master

A. Limits involving trigonometric functions
B. Derivatives of the six trigonometric functions

Summary and Focus Questions

A. Two very important limits involving sine and cosine are:
$$\lim_{x \to 0} \frac{\sin x}{x} = 1 \text{ and } \lim_{x \to 0} \frac{\cos x - 1}{x} = 0$$
where x is the radian measure of an angle.

Many other limits involving trigonometric functions can be solved when transformed by algebra and limit laws into combinations of these two limits.

Example:
Evaluate $\lim_{t \to 0} \frac{\sin 3t}{4t}$.

Since this resembles $\frac{\sin x}{x}$, let $x = 3t$. Then as $t \to 0$, $x \to 0$ and $\lim_{t \to 0} \frac{\sin 3t}{4t}$

$= \lim_{t \to 0} \left(\frac{3}{4}\right) \frac{\sin 3t}{3t} = \lim_{x \to 0} \left(\frac{3}{4}\right) \frac{\sin x}{x} = \lim\left(\frac{3}{4}\right) 1 = \frac{3}{4}$.

1) $\lim_{x \to 0} \frac{1 - \cos x}{x} = \underline{}$.

0.

2) Find $\lim_{x \to 0} \frac{1}{x \cot x} = \underline{}$.

$\frac{1}{\cos x} = \frac{\sin x}{x \cos x} = \frac{\sin x}{x} \cdot \frac{1}{\cos x}$

$\frac{1}{x \cot x} = \frac{1}{x \frac{\cos x}{\sin x}} = \frac{\sin x}{x \cos x} = \frac{\sin x}{x} \cdot \frac{1}{\cos x}$.

Then, $\lim_{x \to 0} \frac{\sin x}{x} \cdot \frac{1}{\cos x} = 1 \cdot \frac{1}{1} = 1$.

3) Find $\lim_{x \to 0} \frac{x^2}{2 \sin x}$.

$\frac{x}{2} \cdot \frac{x}{\sin x}$ $1 + \frac{1}{1} = 1$

$0 \cdot 1 = 0$

$\lim_{x \to 0} \frac{x^2}{2 \sin x} = \lim_{x \to 0} \frac{x}{2} \cdot \frac{x}{\sin x} = \lim_{x \to 0} \frac{x}{2} \cdot \lim_{x \to 0} \frac{x}{\sin x}$
$= (0)(1) = 0$.

52

B. These six differentiation formulas <u>must</u> be memorized.

$$\frac{d}{dx} \sin x = \cos x \qquad\qquad\qquad \frac{d}{dx} \cos x = -\sin x$$

$$\frac{d}{dx} \tan x = \sec^2 x \qquad\qquad\qquad \frac{d}{dx} \cot x = -\csc^2 x$$

$$\frac{d}{dx} \sec x = (\sec x)(\tan x) \qquad\qquad \frac{d}{dx} \csc x = -(\csc x)(\cot x)$$

4) Find y' for each:

a) $y = \sin x - \cos x$

$y' = \cos x + \sin x$

$$y' = \cos x + \sin x.$$

b) $y = \frac{\tan x}{x+1}$

$y' = \frac{(x+1)(\sec^2 x) - (\tan x)(1)}{(x+1)^2}$

$\frac{x \sec^2 x + \sec^2 x - \tan x}{(x+1)^2}$

$$y' = \frac{(x+1)(\sec^2 x) - (\tan x)(1)}{(x+1)^2}$$
$$= \frac{x \sec^2 x + \sec^2 x - \tan x}{(x+1)^2}.$$

c) $y = \sin \frac{\pi}{4}$

$y' = \cos \frac{\pi}{4} = 0$

$$y' = 0, \text{ since } y \text{ is a constant.}$$

d) $y = x^3 \sin x$

$3x^2(\sin x) + \cos x (x^3)$

$3x^2 \sin x + x^3 \cos x$

$$y' = x^3(\cos x) + (\sin x)3x^2$$
$$= x^3 \cos x + 3x^2 \sin x.$$

e) $y = x^2 + 2x \cos x$

$2x + (\cos x)(2) + 2x(-\sin x)$

$2x + 2\cos x - 2x \sin x$

$2(x + \cos x - x \sin x)$

$$y' = 2x + 2x(-\sin x) + 2 \cos x$$
$$= 2(x - x \sin x + \cos x).$$

f) $y = \frac{x}{\sec x + 1}$

$y' = \frac{(\sec x + 1) - x(\sec x)(\tan x)}{(\sec x + 1)^2}$

$$y' = \frac{(\sec x + 1)1 - x(\sec x \tan x)}{(\sec x + 1)^2}$$
$$= \frac{\sec x + 1 - x \sec x \tan x}{(\sec x + 1)^2}$$

g) $y = \frac{x}{\cot x}$

$\frac{x}{\frac{1}{\tan x}}$ $\quad y = x \tan x$

$= x \sec^2 x + \tan x (1)$

First rewrite as $y = x \tan x$.

$$y' = x \sec^2 x + \tan x$$

The Chain Rule

Concepts to Master

Chain Rule for computing the derivative of a composition

Summary and Focus Questions

If $y = f(u)$ and $u = g(x)$ (and thus $y = f \circ g(x)$) then the <u>Chain Rule</u> for computing the derivative of such a composition is:
$$\frac{dy}{dx} = \frac{dy}{du} \cdot \frac{du}{dx}$$
or equivalently $[f(g(x))]' = f'(g(x)) \cdot g'(x)$.

For example, if $y = (x^2 + 6x + 1)^4$, we recognize $f(x) = x^4$ and $g(x) = x^2 + 6x + 1$ as components such that $y = f((g(x)))$.
Thus, $y' = \underbrace{4(x^2 + 6x + 1)^3}_{f'(g(x))} \cdot \underbrace{(2x + 6)}_{g'(x)}$.

The key to correct usage of the Chain Rule comes from first recognizing that the function to be differentiated is a composite and then determining the components. It is often possible to break down a composite in more than one way. You should use components that are easily differentiated. Composite functions often occur as expressions raised to a power, or as some function (such as log, sin, etc.) of an expression.

<u>Examples:</u>   In each of these, y can be written as $f \circ g(x)$:

$y = (x + 3)^4$

$y = f(u) = u^4$ (the "outer" function)
$u = g(x) = x + 3$ (the "inner" function)

Thus, $y' = 4(x + 3)^3 \cdot 1 = 4(x + 3)^3$

$$y = \cos x^2 \qquad\qquad\qquad y = f(u) = \cos u,$$
$$u = g(x) = x^2$$
Thus, $y' = -(\sin x^2) \cdot 2x = -2x \sin x^2$

$$y = \sqrt{1 + \tfrac{1}{x}} \qquad\qquad\qquad y = f(u) = \sqrt{u},$$
$$u = g(x) = 1 + \tfrac{1}{x} = 1 + x^{-1}$$
Thus, $y' = \tfrac{1}{2}\left(1 + \tfrac{1}{x}\right)^{-1/2} \cdot (0 + (-1)x^{-2}) = \dfrac{-1}{2x^2\sqrt{1+\frac{1}{x}}}$

Sometimes more than two functions are necessary, as in:

$$y = (3 + (x^3 - 2x)^5)^8 \qquad y = f(u) = u^8 \text{ (the "outer" function)}$$
$$u = g(v) = 3 + v^5 \text{ (the "middle" function)}$$
$$v = h(x) = x^3 - 2x \text{ (the "inner" function)}$$

Here $y = f \circ g \circ h(x)$ and $\frac{dy}{dx} = 8(3 + (x^3 - 2x)^5)^7 \underbrace{(5(x^3 - 2x)^4)}\underbrace{(3x^2 - 2)}$
$$\underbrace{}_{\frac{dy}{du}} \quad \underbrace{}_{\frac{du}{dv}} \quad \underbrace{}_{\frac{dv}{dx}}$$

1) Suppose $h(x) = f(g(x))$, $f'(7) = 3$,
    $g(4) = 7$, and $g'(4) = 5$. Find $h'(4)$.

    $h'(4) = f(g(4)) \cdot g'(4) = f'(7) \cdot 5$
    $3 \cdot 5 = 15 = 3$

2) Find $\frac{dy}{dx}$ for:

    a) $y = \sqrt{x^3 + 6x}$
    $(x^3 + 6x)^{\frac{1}{2}} \quad \frac{1}{2}(x^3 + 6x)(3x^2 + 6)$
    $\frac{3x + 6}{2(x^3 + 6x)^{\frac{1}{2}}} " = \frac{3x^2 + 6}{2\sqrt{x^2 + 6x}}$

    b) $y = \sec x^2$
    $(\sec x)(\tan x)(2x)$

    c) $y = \sec^2 x$
    $2\sec x(\sec x \tan x) = 2\sec^2 x \tan x$

    d) $y = \cos^3 x^2$
    $3\cos^2 x^2(-\sin^2 x)(2x)$
    $-6x\cos^2 x \cdot \sin^2 x^2$

$h'(4) = f'(g(4)) \cdot g'(4) = f(7) \cdot 5$
$\qquad = 3 \cdot 5 = 15 = 3$

Since $y = (x^3 + 6x)^{1/2}$
$y' = \tfrac{1}{2}(x^3 + 6x)^{-1/2}(3x^2 + 6) = \frac{3x^2 + 6}{2\sqrt{x^3 + 6x}}$

$y' = (\sec x^2)(\tan x^2)(2x)$.

Since $y = [\sec x]^2$,
$y' = 2[\sec x]^1(\sec x \tan x) = 2\sec^2 x \tan x$.

Since $y = [\cos x^2]^3$,
$y' = 3[\cos x^2]^2(-\sin x^2(2x))$
$\qquad = -6x\cos^2 x^2 \sin x^2$.

e) $y = \sin 2x \cos 3x$

$\sin 2x (-\sin 3x)(3) +$

$(\cos 3x)(\cos 2x)(2)$

$-3\sin 2x \sin 3x + 2\cos 3x \cos 2x$

y is a product of $\sin 2x$ and $\cos 3x$, so
$$y' = (\sin 2x)[-\sin 3x \cdot 3]$$
$$+ (\cos 3x)[\cos 2x \cdot 2]$$
$$= -3\sin 2x \sin 3x + 2\cos 3x \cos 2x.$$

f) $y = \sin(\sec x)$

$\cos(\sec x)(\sec x \tan x)$

$$y' = \cos(\sec x)(\sec x \tan x).$$

g) $y = \tan(x^2 + 1)^4$

$\sec^2 (x^2 + 1)^4 (4(x^2 + 1)(2x)) -$

$8x \sec^2 (x^2 + 1)^4 (x^2 + 1)$

$$y' = \sec^2(x^2 + 1)^4 (4(x^2 + 1)^3 (2x))$$
$$= 8x(x^2 + 1)^3 \sec^2(x^2 + 1)^4.$$

h) $y = \frac{1}{(x^2 - 1)^4}$ $(x^2 - 1)^{-4}$

$-4(x^2 - 1)^{-5}(2x)$

$\frac{-8x}{(x^2 - 1)^5}$

Since $y = (x^2 - 1)^{-4}$,
$$y' = -4(x^2 - 1)^{-5}(2x) = \frac{-8x}{(x^2 - 1)^5}$$

i) $y = x^3 \sqrt{x^2 + 1}$ $(x^2 + 1)^{\frac{1}{2}}$

$x^3 (\frac{1}{2}(x^2 + 1)^{-\frac{1}{2}})(2x) + 3x^2 (x^2 + 1)^{\frac{1}{2}}$

$x^4 (x^2 + 1)^{-\frac{1}{2}} + 3x^2 (x^2 + 1)^{\frac{1}{2}}$

$\frac{x^4}{(x^2 + 1)^{\frac{1}{2}}} + 3x^2 (x^2 + 1)^{\frac{1}{2}}$

$\frac{x^4}{\sqrt{x^2 + 1}} + 3x^2 \sqrt{x^2 + 1}$?

Since $y = x^3 (x^2 + 1)^{1/2}$,
$$y' = x^3 \left(\frac{1}{2}(x^2 + 1)^{-1/2} \cdot 2x \right)$$
$$+ (x^2 + 1)^{1/2} \cdot 3x^2$$
$$= \frac{x^4}{\sqrt{x^2 + 1}} + 3x^2 \sqrt{x^2 + 1}$$
$$= \frac{4x^4 + 3x^2}{\sqrt{x^2 + 1}}$$

Implicit Differentiation

Concepts to Master

A. Defining a functional relationship implicitly
B. Finding $\frac{dy}{dx}$ where the function y of x is given implicitly

Summary and Focus Questions

A. If y is a function of x defined by an equation not of the form $y = f(x)$, we say y as a function of x is defined <u>implicitly</u> by the equation.

<u>Example</u>:
$x + \sqrt{y} = 1$ defines y as a function of x. If we solve for y (and remember that y must be nonnegative) we get $y = (1 - x)^2$ with domain $(-\infty, 1]$.

The graph of a relation is not necessarily the graph of a function. A portion of the graph may be the graph of a function. That function is defined implicitly be the equation.

For example, the $x = y^2$ is a parabola opening to the right. It is not the graph of a function but the top half is the graph of the function $y = \sqrt{x}$ and bottom half is the graph of $y = -\sqrt{x}$. Both functions are defined implicitly bty $x = y^2$.

1) Sketch a graph of $\frac{x^2}{9} - \frac{y^2}{16} = 1$ and determine explicit forms for the functions it defines.

We recognize the graph of the equation as a hyperbola opening left and right with asymptotes $y = \pm \frac{4}{3}x$.

$\frac{1}{9}x^2 - \frac{1}{16}y^2 = 1$

$\frac{1}{9}(2x) - \frac{1}{16}2yy' = 0$

$\frac{2x}{9} - \frac{2yy'}{168} = 0$

$\frac{2x}{9} - \frac{yy'}{8} = 0$

$\frac{2x}{9} = \frac{yy'}{8}$

$\frac{16x}{9} = yy'$

$\frac{4}{3}x = y$

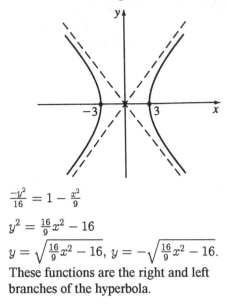

$$\frac{-y^2}{16} = 1 - \frac{x^2}{9}$$

$$y^2 = \frac{16}{9}x^2 - 16$$

$$y = \sqrt{\frac{16}{9}x^2 - 16}, \; y = -\sqrt{\frac{16}{9}x^2 - 16}.$$

These functions are the right and left branches of the hyperbola.

2) The graph below is $x^{2/3} + y^{2/3} = 1$, which implicitly defines two functions y of x. Find the functions explicitly.

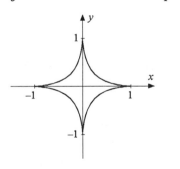

Solve for y:
$$x^{2/3} + y^{2/3} = 1$$
$$y^{2/3} = 1 - x^{2/3}$$
$$y^2 = (1 - x^{2/3})^3$$
$$y = (1 - x^{2/3})^{3/2} \text{ (the top half) and}$$
$$y = -(1 - x^{2/3})^{3/2} \text{ (the bottom half).}$$

B. Given an equation that defines y as an implicit function of x, $\frac{dy}{dx}$ may be found by implicit differentiation:

Take the derivative with respect to x of both sides of the equation, then solve the result for $\frac{dy}{dx}$.

The Chain Rule is used for terms involving y.

3) Find $\frac{dy}{dx}$ implicitly:

a) $x^2 y^3 = 2x + 1$

$x^2(3y^2 y') + 2x(y^3) = 2$

$x^2(3y^2 y') = 2 - 2xy^3$

$3y^2 y' = \frac{2 - 2xy^3}{x^2}$

Differentiate with respect to x (using the Product Rule because both x^2 and y^3 are functions of x): $x^2[3y^2 y'] + y^3[2x] = 2$.
Solve for y': $3x^2 y^2 y' = 2 - 2xy^3$
$y' = \frac{2 - 2xy^3}{3x^2 y^2}$.

b) $3x^2 - 5xy + y^2 = 10$

$6x - 5x(y') + 5y + 2yy' = 0$

$6x - 5xy' + 5y + 2yy' = 0$

$6x + 5y = 5xy' + 2yy'$

$\frac{6x + 5y}{5x + 2y} = y'$

Differentiate:
$6x - (5x(y') + y \cdot 5) + 2y \cdot y' = 0$.
Solve for y': $-5xy' + 2yy' = 5y - 6x$
$(2y - 5x)y' = 5y - 6x$
$y' = \frac{5y - 6x}{2y - 5x}$.

4) Find the slope of the tangent line to the curve defined by
$x^2 + 2xy - y^2 = 1$ at the point $(5, 2)$.

$2x + (2xy' + 2y) - 2yy' = 0$

$2x + 2xy' + 2y - 2yy' =$
$2x + 2y = 2yy' - 2xy'$ $y'(2y - 2x)$

$\frac{2x + 2y}{2y - 2x} \cdot \frac{2(x+y)}{2(y-x)} = \frac{x+y}{y-x}$ $\frac{5+2}{2-5} = \frac{7}{-3}$

The slope is $\frac{dy}{dx}$ at $(5, 2)$. First find $\frac{dy}{dx}$ implicitly:
$2x + (2xy' + y \cdot 2) - 2yy' = 0$
$2y'(x - y) = -2(x + y)$
$y' = -\frac{x+y}{x-y} = \frac{x+y}{y-x}$.
Now use $x = 5$, $y = 2$:
$\frac{dy}{dx} = \frac{5+2}{2-5} = \frac{7}{-3} = -\frac{7}{3}$.

5) For $x^2 y = 1$ find y' both explicitly and implicitly.

Imp

$x^2 y' + 2xy = 0$

$x^2 y' = -2xy$

$y' = \frac{-2xy}{x^2} = \frac{-2y}{x}$

Exp

$x^2 y = 1$

$y = x^{-2}$

$y' = -2x^{-3}$

Explicitly:
From $x^2 y = 1$, $y = x^{-2}$ so $y' = -2x^{-3}$.
Implicitly:
$x^2 y' + 2xy = 0$
$x^2 y' = -2xy$
$y' = -\frac{2y}{x}$
These are the same because from $y = x^{-2}$
$y' = \frac{-2y}{x} = \frac{-2x^{-2}}{x} = -2x^{-3}$.

Higher Derivatives

Concepts to Master

A. Higher derivatives; Higher derivative notations
B. Acceleration as the derivative of velocity
C. Relationships between f, f', and f''

Summary and Focus Questions

A. The second derivative of $y = f(x)$, denoted y'', is the derivative of y'. The third derivative, y''', is the derivative of y''. The nth derivative is denoted $y^{(n)}$. For $y = x^6$,

$$y' = 6x^5$$
$$y'' = 30x^4$$
$$y''' = 120x^3$$
$$y^{(4)} = 360x^2.$$

Each of these notations may be used for the nth derivative of y with respect to x:

$$y^{(n)}, \quad f^{(n)}(x), \quad \frac{d^n y}{dx^n}, \quad D^n f(x), \quad D^n y$$

1) Find $y^{(2)}$ for:

a) $y = 5x^3 + 4x^2 + 6x + 3$
$y' = 15x^2 + 8x + 6$
$y'' = 30x + 8$

$y' = 15x^2 + 8x + 6$,
so $y'' = = 30x + 8$.

b) $y = \sqrt{x^2 + 1}$ $\quad (x^2+1)^{\frac{1}{2}}$
$y' = \frac{1}{2}(x^2+1)^{-\frac{1}{2}}(2x)$
$\quad x(x^2+1)^{-\frac{1}{2}}$

$y'' = x(\ominus\frac{1}{2}(x^2+1)^{-\frac{3}{2}}(2x)) +$
$\quad (1)(x^2+1)^{\frac{1}{2}}$
$-x^2(x^2+1)^{-\frac{3}{2}} + (x^2+1)^{\frac{1}{2}}$

$y = (x^2 + 1)^{1/2}$
$y' = \frac{1}{2}(x^2 + 1)^{-1/2}(2x) = x \cdot (x^2 + 1)^{-1/2}$
Thus, $y'' =$
$x\left[-\frac{1}{2}(x^2 + 1)^{-3/2}(2x)\right] + (x^2 + 1)^{-1/2}(1)$
$\quad = -x^2(x^2 + 1)^{-3/2} + (x^2 + 1)^{-1/2}.$

2) True or False:
The notation for the sixth derivative of
$y = f(x)$ is $\frac{dy^6}{d^6x}$.

False

False. It is $\frac{d^6y}{dx^6}$.

3) Find $f^{(4)}(x)$, where $f(x) = \frac{x^{10}}{90} + \frac{x^5}{60}$.

$f(x) \frac{1}{90} x^{10} + \frac{1}{60} x^5 = \frac{1}{9} x^9 + \frac{1}{12} x^4$

$f''(x) = x^8 + \frac{1}{3} x^3$

$f'''(x) = 8x^7 + x^2$ $f''''= 56x^6 + 2x$

$f'(x) = \frac{x^9}{9} + \frac{x^4}{12}$
$f^{(2)}(x) = x^8 + \frac{x^3}{3}$
$f^{(3)}(x) = 8x^7 + x^2$
$f^{(4)}(x) = 56x^6 + 2x$

4) Find a formula for $f^{(n)}(x)$, where
$f(x) = \frac{1}{x^2}$.

$f'(x) = x^{-2} = -2x^{-3}$ $(-2)(-3)x^{-3}$

$f''(x) = 6x^{-4}$ $(-2)(-3)(-4)x^{-4}$

$f'''(x) = -24x^{-5}$ $-(n+1)x^{-(n+2)}$

$f''''(x) = 120x^{-6}$ $\frac{-1^n(n+1)!}{x^{n+2}}$

$f(x) = x^{-2}$
$f^{(1)}(x) = -2x^{-3}$
$f^{(2)}(x) = (-2)(-3)x^{-4}$
$f^{(3)}(x) = (-2)(-3)(-4)x^{-5}$.
The pattern is
$f^{(n)}(x) = (-2)(-3)\ldots(-(n+1))x^{-(n+2)}$
$= \frac{(-1)^n(n+1)!}{x^{n+2}}$.

5) If $y = f(x)$ is a polynomial of degree
n, what is $y^{(n+1)}$.

$y^{(n+1)} = 0$ y' has degree n 1, y'' has
degree $n-2$, and so on. $y^{(n)}$ has degree
zero (is a constant). Thus, $y^{(n+1)} = 0$.

B. The rate of change of velocity is called acceleration. Thus, if $y = f(x)$ is a
position function, then y'' measures the instantaneous acceleration of the
particle.

6) Find the velocity and acceleration after
2 seconds of a particle whose position
(in meters) after x seconds is
$s = 3t^3 + 6t^2 + t + 1$.

$s' = 9t^2 + 12t + 1$ $s'' = 18t + 12$

$= 9(2)^2 + 12(2) + 1$ $= 18(2) + 12$

$= 9(4) + 24 + 1$ $= 36 + 12$

$\begin{array}{c} 36 \\ 24 \\ 1 \end{array}$ $= 48 \, m/s^2$

$61 \, m/s$

$s' = 9t^2 + 12t + 1$. At $t = 2$, the velocity
is $s' = 9(2)^2 + 12(2) + 1 = 61$ m/sec.
$s'' = 18t + 12$. At $t = 2$, the acceleration is
$s'' = 18(2) + 12 = 48$ m/sec^2

7) Find the acceleration at time $\frac{\pi}{6}$ for a particle whose position in meters at time t seconds is $s = \cos 2t$.

$s' = 2\sin 2t$ $4\cos 2\left(\frac{\pi}{6}\right)$

$s'' = -4\cos 2t$ $4\cos\left(\frac{\pi}{3}\right) = 4\left(\frac{1}{2}\right) = 2$

$s = \cos 2t, \; s' = -2\sin 2t,$
$s'' = -4\cos 2t.$ At $t = \frac{\pi}{6},$
$s'' = -4\cos\frac{\pi}{3} = -4\left(\frac{1}{2}\right) = -2$ m/sec^2

C. The graph of $f''(x)$ has the same relationship to the graph of $f'(x)$ as the graph of $f'(x)$ does to the graph of $f(x)$ — $f'(x)$ is the slope of the tangent line to $y = f'(x)$.

8) Use your best judgement to determine the graph of $y = f''(x)$ given this graph of $y = f(x)$.

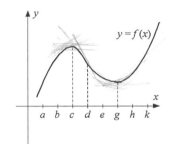

From the slopes of tangents to $y = f(x)$ the graph of $y = f'(x)$ is:

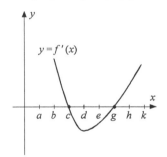

From the slopes of tangents to this graph of $y = f'(x)$ the graph of $y = f''(x)$ is:

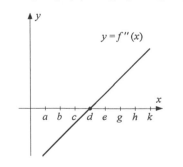

Related Rates

Solutions of related rate problems · yeah, okay.

Summary and Focus Questions

The procedure to solve a related rate problem is:

1) Illustrate the problem with a picture if possible.
2) Identify and label all fixed quantities with constants and quantities that are functions of time with variables. Any rates that are given in the problem are the rates of change of these variables. The unknown rate will be the rate of change of another one of these variable(s).
3) Find an equation that relates the constants, the variables whose rates are given, and the variable whose rate is desired.
4) Differentiate the equation in (3) with respect to time.
5) Substitute all known quantities in the result of (4) and solve for the unknown rate.

Step (3) is often the hardest and will call upon your skills in remembering relationships from geometry, trigonometry, ...

1) An observer, 300 meters from the launch pad of a rocket, watches it ascend vertically at 60 m/sec. Find the rate of change of the distance between the rocket and the observer when the rocket is 400 meters high.

(1)

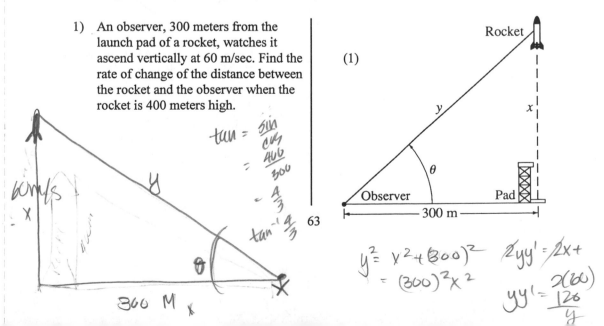

$\tan = \frac{\sin}{\cos}$

$= \frac{400}{300}$

$= \frac{4}{3}$

$\tan^{-1} \frac{4}{3}$

63

$y^2 = x^2 + (300)^2$

$= (300)^2 x^2$

$2yy' = 2x +$

$yy' = \frac{2(60)}{120}$

$\frac{}{y}$

(2) Let $y =$ distance between rocket and observer. Let $x =$ distance between rocket and pad. $\frac{dx}{dt}$ is known (60); $\frac{dy}{dt}$ is unknown.

(3) Relate y, x, and 300: $y^2 = 300^2 + x^2$.

(4) Use implicit differentiation, remembering y and x are functions of time t:
$$2y\frac{dy}{dt} = 0 + 2x\frac{dx}{dt}$$
$$y\frac{dy}{dt} = x\frac{dx}{dt}.$$

(5) Substitute known values:
$\frac{dx}{dt} = 60$, $x = 400$. The value of y is known when $x = 400$; from
$y^2 = 300^2 + (400)^2$, $y = 500$.
$$y\frac{dy}{dt} = x\frac{dx}{dt}$$
$$500\frac{dy}{dt} = 400(60)$$
$$\frac{dy}{dt} = 48 \text{ m/sec.}$$

2)   A spherical ball has its diameter increasing at 2 inches/sec. How fast is the volume changing when the radius is 10 inches?

(1)

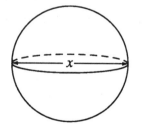

(2) Let x be the diameter, V the volume. We are given $\frac{dx}{dt} = 2$.

(3) Relate V and x:
$V = \frac{4}{3}\pi\left(\frac{x}{2}\right)^3 = \frac{\pi}{6}x^3$.

(4) Differentiate:
$\frac{dV}{dt} = \frac{3\pi}{6}x^2\frac{dx}{dt} = \frac{\pi x^2}{2}\frac{dx}{dt}$.

(5) When the radius is 10, $x = 20$.
$\frac{dV}{dt} = \frac{\pi(20)^2}{2} \cdot 2 = 400\pi$ in³/sec.

$$\frac{4}{3}\pi \frac{x^3}{2^3} = \frac{4\pi x^3}{3 \cdot 8}_2$$

$$\frac{\pi x^3}{6}$$

$$\frac{1}{6} \; 3 x \pi$$

radius-10
diam-20

$$\frac{x^2 \pi}{2}\frac{dx}{dt}$$

$$\frac{(20)^2 \pi}{2} = \frac{400 \pi \cdot 2}{2}$$

$$= 400\pi \text{ in}^3/s$$

3) A light is on top of a 12 ft vertical pole. A 6 ft woman walks away from the pole base at 4 ft/sec. How fast is the angle made by the woman's head, the light, and the pole base changing 2 seconds after she starts walking?

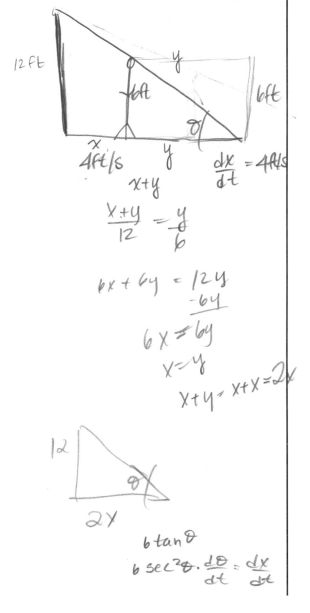

$12ft$

y

$6ft$

$6ft$

x

$4ft/s$ y $\frac{dx}{dt} = 4ft/s$

$x+y$

$\frac{x+y}{12} = \frac{y}{6}$

$6x + 6y = 12y$

$-6y$

$6x \not= 6y$

$x \sim y$

$x+y \sim x+x = 2x$

12

θ

$2x$

$6 \tan \theta$

$6 \sec^2 \theta \cdot \frac{d\theta}{dt} = \frac{dx}{dt}$

(1)

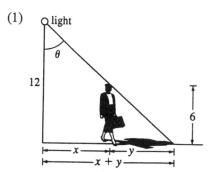

(2) Let θ be the desired angle. Let x be the distance between the pole base and the woman. Let y be the distance from the woman to the tip of her shadow.

(3) We are given $\frac{dx}{dt} (= 4)$ and we must find $\frac{d\theta}{dt}$. Thus, we must relate x and θ. We can relate $x + y$ and θ so we first determine y in terms of x. By similar triangles $\frac{x+y}{12} = \frac{y}{6}$.

Thus, $\qquad 6x + 6y = 12y$

$\qquad\qquad 6x = 6y$, or $x = y$.

Thus, $\qquad x + y = x + x = 2x$.

By trigonometry, we obtain the desired relation between x and θ:

$\tan \theta = \frac{2x}{12}$, or $6 \tan \theta = x$.

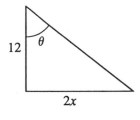

(4) Differentiate the relation:

$6 \sec^2 \theta \cdot \frac{d\theta}{dt} = \frac{dx}{dt}$.

(5) After 2 seconds, $x = 2(4) = 8$ so
$2x = 16$ ft. $\sec \theta = \frac{d}{12}$,
so $\sec^2 \theta = \frac{d^2}{144}$.

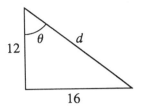

From the figure $12^2 + 16^2 = d^2$ so
$d^2 = 400$ and $\sec^2 \theta = \frac{400}{144}$. Finally
substituting into the result from (4):
$6\left(\frac{400}{144}\right)\frac{d\theta}{dt} = 4$
$\frac{d\theta}{dt} = 2.4$ radians/sec.

4)

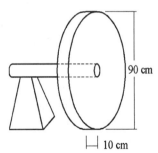

90 cm

10 cm

A 10 cm thick mill wheel is initially 90 cm in diameter and is wearing away (uniformly) at 55 cm^3/hr. How fast is the diameter changing after 20 hours?

(1)

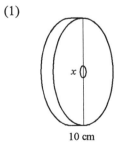

x

10 cm

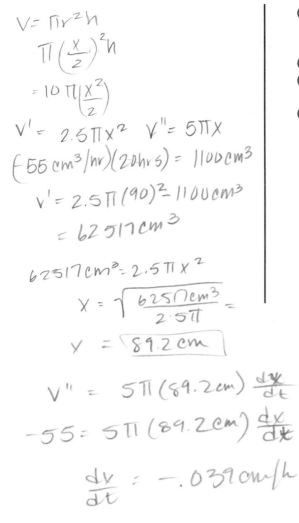

$V = \pi r^2 h$

$\pi \left(\dfrac{x}{2}\right)^2 h$

$= 10 \, \pi \left(\dfrac{x^2}{2}\right)$

$V' = 2.5 \pi x^2 \qquad V'' = 5\pi x$

$\left(55 \, cm^3/hr\right)\left(20 hrs\right) = 1100 \, cm^3$

$V' = 2.5 \pi (90)^2 - 1100 \, cm^3$

$= 62517 \, cm^3$

$62517 \, cm^3 = 2.5 \pi x^2$

$x = \sqrt{\dfrac{62517 \, cm^3}{2.5 \pi}} =$

$x = \boxed{89.2 \, cm}$

$V'' = 5\pi (89.2 \, cm) \dfrac{dx}{dt}$

$-55 = 5\pi (89.2 \, cm) \dfrac{dx}{dt}$

$\dfrac{dv}{dt} = -.039 \, cm/h$

(2) Let V be the volume and x the diameter, functions of time. $\frac{dV}{dt}$ is known (-55), $\frac{dx}{dt}$ is unknown.

(3) $V = 10\pi\left(\frac{x}{2}\right)^2$, so $V = 2.5\pi x^2$.

(4) Differentiate with respect to time: $\frac{dV}{dt} = 5\pi x \cdot \frac{dx}{dt}$.

(5) 20 hours later the amount of material worn away is $(55)20 = 1100 \text{ cm}^3$. The volume after 20 hours is $2.5\pi(90)^2 - 1100 = 62517 \text{ cm}^3$. Thus the diameter x after 20 hours is $62517 = 2.5\pi x^2$
$$x = \sqrt{\frac{62517.25}{2.5\pi}} = 89.2 \text{ cm.}$$
Finally, from $\frac{dV}{dt} = 5\pi x \frac{dx}{dt}$
$-55 = 5\pi(89.2)\frac{dx}{dt}$
$\frac{dx}{dt} = -0.039 \text{ cm/hr.}$

Differentials; Linear and Quadratic Approximations

Concepts to Master

A. Differentials; Approximations with differentials
B. Linear approximation
C. Quadratic approximation

Summary and Focus Questions

A. Let $y = f(x)$ be a differentiable function. The differential dx is an independent variable. The differential dy is defined as: $dy = f'(x)dx$.

Note that dy is a function of both x (because of $f'(x)$) and dx.

If we let $dx = \Delta x$, then for small values of dx, the change in the function (Δy) is approximately the same as the change in the tangent line dy: $dy \approx \Delta y$ for dx small.

This is handy since dy may be easier to calculate than Δy.

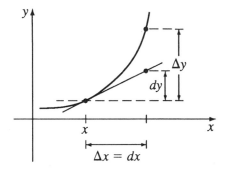

1) True or False: $dy = \Delta y$.

False.

2) Compute dy and Δy for $f(x) = x^2 + 3x$ at $x = 2$ with $\Delta x = dx = 0.1$.

$f'(x) = 2x + 3 \qquad f'(x) \cdot dy$

$= 2(2) + 3 \qquad 7(0.1)$

$f'(x) = 7 \qquad \boxed{= 0.7}$

68

$f(2) = x^2 + 3x$
$2^2 + 3(2)$
$= 4 + 6 = 10$

$f(2) + \Delta x = 2.1$
$f(2.1) = (2.1)^2 + 3(2.1)$
$= 4.2 + 6.3$
$= 10.71$
$- 10$
$\boxed{.71}$

$f(x) = x^2 + 3x.$
① $f' = 2x + 3.$
② At $x = 2$, $f'(2) = 2(2) + 3 = 7.$
③ Thus, $dy = f'(x)dx = f'(2)(0.1)$
$= 7(0.1) = \boxed{0.7.}$
⓪ At $x = 2$, $y = f(2) = 2^2 + 3(2) = 10.$
④ At $x = 2 + \Delta x = 2 + 0.1 = 2.1,$
$y = f(2.1) = (2.1)^2 + 3(2.1) = 10.71.$
Thus, $\Delta y = f(2.1) - f(2)$
$= 10.71 - 10 = 0.71.$
Note that $dy = 0.7$ and $\Delta y = 0.71$ are
quite close but dy is easier to calculate.

3) A circle has a radius of 20 cm with a
possible error of 0.2 cm. Use
differentials to estimate the maximum
error in the area of the circle.

$A = \pi r^2$
$A' = 2\pi r (dr)$
$2\pi (20)(0.2) = 0.8\pi$

Let $x =$ radius of circle and $A =$ area of
circle. We are given $\Delta x = dx = 0.2$ for
$x = 20$. The error is ΔA which we
estimate with dA: $A = \pi x^2$, so
$dA = 2\pi x \, dx = 2\pi(20) \cdot (0.02) = 0.8\pi.$

B. Let $L(x)$ be the equation of the
tangent line to $y = f(x)$ at $x = a$:
$L(x) = f(a) + f'(a)(x - a)$
Both $L(x)$ and $f(x)$ have the same
first derivative so they both "head in
the same direction" $(f'(a))$ at
$x = a$.
For x near a, $L(x)$ may be used as a
linear approximation to $f(x)$:
$L(x) \approx f(x).$

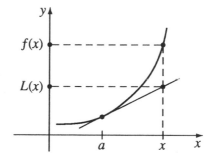

4) Find the linear approximation to
$f(x) = 5x^3 + 6x$ at $x = 2$.
$= 5(2)^3 + 6(2)$
$= 5(8) + 12$
$= 52$
$f'(x) = 15x^2 + 6$
$= 15(2)^2 + 6$
$= 15(4) + 6$
$= 60 + 6 = 66$
$52 + 66(x - 2)$

What we are asked to do is find the
equation of the tangent line:
$f(2) = 5(2)^3 + 6(2) = 52.$
$f'(x) = 15x^2 + 6$, so
$f'(2) = 15(2)^2 + 6 = 66.$
Thus, $L(x) = 52 + 66(x - 2).$

5) Approximate $f(1.98)$ for the function in question 4.

$55 + 66(1.98-2) = 50.68$

We use $f(1.98) \approx L(1.98)$.
$L(1.98) = 52 + 66(1.98 - 2) = 50.68$.
(*Note*: $f(1.98) = 50.69196$, so $L(1.98)$ is rather close and a lot easier to calculate.)

6) Approximate $\sqrt{66}$ using a linear approximation.

Choose $f(x) = \sqrt{x}$ and $a = 64$ (64 is near 66). Then $f(64) = 8$, $f'(x) = \frac{1}{2\sqrt{x}}$, and
$f'(64) = \frac{1}{2\sqrt{64}} = \frac{1}{6}$.
The linear approximation is:
$L(x) = 8 + \frac{1}{16}(x - 64)$.
At $x = 66$,
$L(66) = 8 + \frac{1}{16}(66 - 64) = 8.125$.
(Note: $\sqrt{66} \approx 8.1240384$).

7) Find the linear approximation of $f(x) = \sin x$ at $a = 0$.

$f(x) = \sin x$
$f(0) = \sin 0 = 0$.
$f'(x) = \cos x$
$f'(0) = \cos 0 = 1$.
So $L(x) = f(0) + f'(0)(x - 0)$
$\qquad = 0 + 1(x - 0) = x$.
Thus, $\sin x \approx x$, for x near 0.
(Note: This is an approximation that you may see in other courses such as physics.)

C. Let $P(x)$ be the equation of a quadratic which has the same first and second derivative as f at $x = a$. Then $P(x)$ has the form :
$$P(x) = f(a) + f'(a)(x - a) + \frac{f''(a)}{2}(x - a)^2$$

For x near a, $P(x)$ may be used as a quadratic approximation to $f(x)$:
$P(x) \approx f(x)$

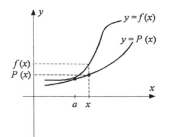

8) Find the quadratic approximation to $f(x) = 5x^3 + 6x$ at $x = 2$.

$f(2) = 5(2)^3 + 6(2) = 52.$
$f'(x) = 15x^2 + 6$ so
$f'(2) = 15(2)^2 + 6 = 66.$
$f''(x) = 30x$ so $f''(2) = 30(2) = 60.$
Thus,
$P(x) = 52 + 66(x - 2) + \frac{60}{2}(x - 2)^2$ or
$P(x) = 52 + 66(x - 2) + 30(x - 2)^2$

9) Approximate $f(1.98)$ for the function in question 8.

We use $f(1.98) \approx P(1.98)$.
$P(1.98) = 52 + 66(1.98 - 2)$
$\qquad + 30(1.98 - 2)^2 = 50.692.$
Note that this quadratic approximation is better than the linear approximation found in question 5.

10) Find the quadratic approximation of $f(x) = \tan x$ at $x = \frac{\pi}{4}$.

$f(x) = \tan x, \ f\left(\frac{\pi}{4}\right) = \tan \frac{\pi}{4} = 1.$
$f'(x) = \sec^2 x,$
$f'\left(\frac{\pi}{4}\right) = \sec^2 \frac{\pi}{2} = \left(\sqrt{2}\right)^2 = 2.$
$f''(x) = 2(\sec x) \sec x \tan x$
$\qquad = 2\sec^2 x \tan x,$
$f''\left(\frac{\pi}{4}\right) = 2\sec^2 \frac{\pi}{4} \tan \frac{\pi}{4} = 2 \cdot 2 \cdot 1 = 4$
$P(x) = 1 + 2\left(x - \frac{\pi}{4}\right) + 4\left(x - \frac{\pi}{4}\right)^2$

Newton's Method

Concepts to Master

Approximations of solutions to $f(x) = 0$ using Newton's Method

Summary and Focus Questions

Let f be a continuously differentiable function on an open interval with a real root. If x_1 is an estimate of the root, then

$$x_2 = x_1 - \frac{f(x_1)}{f'(x_1)}$$

is often (but not always) a closer approximation to the root. The point x_2 is where the tangent line to f at x_1 crosses the x-axis. The process of

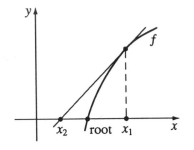

obtaining x_2 may be repeated using x_2 in the role of x_1 to obtain another approximation x_3. In this way we can generate a sequence of approximations x_1, x_2, x_3, ... which can approach the value of the root. In general, to obtain a desired degree of accuracy, repeat Newton's Method until the difference between successive x_i is within that accuracy.

1) Indicate the point x_2 determined by Newton's Method in each:

a)

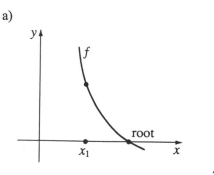

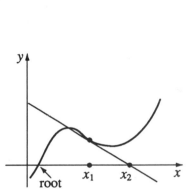

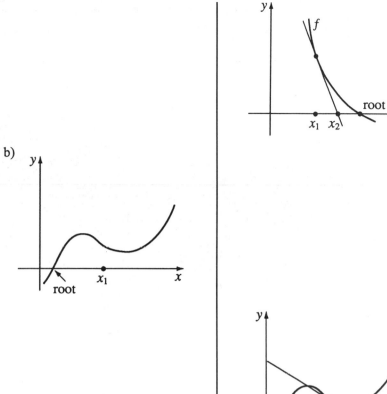

b)

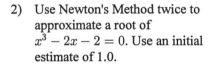

(Note: There is no guarantee that x_2 will be a better estimate than x_1.)

2) Use Newton's Method twice to approximate a root of $x^3 - 2x - 2 = 0$. Use an initial estimate of 1.0.

$f(x) = x^3 - 2x - 2, \ f'(x) = 3x^2 - 2$

$x_1 = 1.0$, so

$x_2 = 1.0 - \frac{f(1.0)}{f'(1.0)} = 1.0 - \frac{-3.0}{1.0} = 4.0.$

$x_3 = 4.0 - \frac{f(4.0)}{f'(4.0)} = 4.0 - \frac{54}{46} \approx 2.826.$

3

The Mean Value Theorem & Curve Sketching

Cartoons courtesy of Sidney Harris. Used by permission.

Maximum and Minimum Values

Concepts to Master

A. Absolute maxima, minima, extrema; Relative (local) maxima, minima, extrema; The Extreme Value Theorem
B. Critical Numbers; Fermat's Theorem about local extrema; Absolute extrema of a continuous function on a closed interval

Summary and Focus Questions

A. Let f be a function with domain D. f has an <u>absolute maximum</u> at c (and $f(c)$ is the <u>maximum value</u>) means $f(x) \leq f(c)$ for all $x \in D$. Thus, a highest point on the graph of f occurs at $(c, f(c))$.

f has an <u>absolute minimum</u> at c means $f(x) \geq f(c)$ for all $x \in D$, so $(c, f(c))$ is a lowest point on the graph.

The <u>extreme values</u> of f are the maximum of f (if there is one) and the minimum of f (if there is one). Extreme values are important in applications (viz., maximum profit, minimum force, etc.). The following three graphs from left to right have two, one, and no extreme values, respectively.

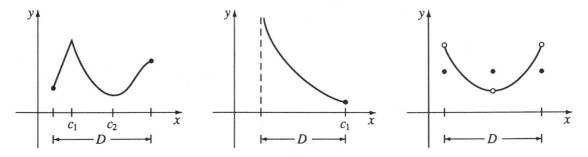

The <u>Extreme Value Theorem</u> says that the <u>extreme values must always exist</u> for a <u>continuous function whose domain is a closed interval</u>.

If there is some open interval I containing c such that $f(c) \geq f(x)$ for all $x \in I$, then f has a <u>local (or relative) maximum</u> at c. In case $f(c) \leq f(x)$ for all $x \in I$, we say f has a <u>local (or relative) minimum</u> at c.

The following graph is a function with domain $[a, b]$ with four local maxima at $x = c, p, r, t$ and five local minima at $x = a, d, q, s, b$.

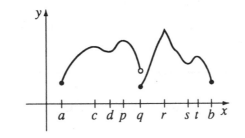

1) If $f(x) \geq f(c)$ for all x in the domain of the function f, then f has an _____ at c.

    Absolute minimum.

2) True or False:
    A function may have more than one absolute minimum value.

    False. (However, this minimum could occur at more than one point.)

3) Where does the absolute minimum value of $f(x) = 9 - x^2$ for $x \in [-2, 2]$ occur?

    $f(x) = 9 - x^2$ with domain $[-2, 2]$ has a minimum value of 5 which occurs at both $x = 2$ and $x = -2$.

4) Does the Extreme Value Theorem guarantee that the extreme values exist for the function graphed?

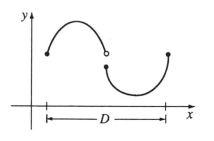

No. The function is not continuous on D. For this function, the extreme values exist, but their existence is not guaranteed by the Extreme Value Theorem.

5) True or False:
   If f has an absolute minimum at $x = c$ then f has a local minimum at $x = c$.

True.

6) How many local maxima and minima does the following function with domain (a, b) have?

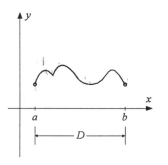

3 local maxima and 4 local minima.

B. A <u>critical number</u> c is a number in the domain of f for which either $f'(c) = 0$ or $f'(c)$ does not exist.

Determining where $f'(x) = 0$ is a matter of solving the equation $f'(x) = 0$. The method of solution depends greatly on the nature of the function. Factoring will sometimes work as in this example.

Example:
Solve $f'(x) = 0$ where $f(x) = 3x^4 + 20x^3 - 36x^2 + 7$.
$f'(x) = 12x^3 + 60x^2 - 72x = 12x(x-1)(x+6)$. So $f'(x) = 0$ at $x = 0$, $x = 1$, $x = -6$.

Determining where $f'(x)$ does not exist is often (but not always!) a matter of finding points where a denominator is 0 or where an even root of a negative number occurs.

Examples:
Find all points in the domain of f at which $f'(x)$ does not exist.
   a) $f(x) = 9\sqrt{x} + x^{3/2}$. Here the domain of f is $[0, \infty]$,
       $f'(x) = \frac{9}{2\sqrt{x}} + \frac{3\sqrt{x}}{2}$ and $f'(x)$ does not exist for $x = 0$.

   b) $f(x) = \sqrt{4-x}$. The domain of f is $(-\infty, 4]$ and $f'(x) = \frac{-1}{2\sqrt{4-x}}$
       does not exist at $x = 4$.

<u>Fermat's Theorem</u> states that if f has a local maximum or minimum at c and $f'(c)$ exists, then $f'(c) = 0$. In other words, if $f(c)$ is a local extremum for f, then c is a critical number of f. The converse is false - a critical number does not have to be where a local maximum nor local minimum occurs.

The extreme values of a continuous function f on a closed interval $[a, b]$ always exist; they occur either at a, at b, or at a critical number of f in (a, b).

7) Suppose $f'(c) = 0$. Is c a local
     maximum or minimum?

Not necessarily; for example, $f(x) = x^3$ with $c = 0$ has $f'(c) = 0$ but $c = 0$ is not a local extremum of $f(x) = x^3$.

8) Find the extreme values of
$f(x) = \sqrt{10x - x^2}$ on $[2, 10]$.

Since f is continuous on $[2, 10]$, the extreme values occur at 2, 10, or some critical number between 2 and 10.
$$f(x) = (10x - x^2)^{1/2}$$
$$f'(x) = \tfrac{1}{2}(10x - x^2)^{-1/2}(10 - 2x)$$
$$= \frac{5-x}{\sqrt{10x-x^2}}.$$
On $[2, 10]$, $f'(x) = 0$ at $x = 5$. $f'(x)$ does not exist at $x = 10$. Thus 5 is the only critical number between 2 and 10.
Computing functional values:

x	5	2	10
$f(x)$	5	4	0

The absolute minimum is 0 at $x = 10$ and the absolute maximum is 5 at $x = 5$.

9) Do the extreme values of
$f(x) = 6 + 12x - x^2$ on $(4, 10]$ exist?

f is a polynomial and, therefore, continuous on $(4, 10]$. However, $(4, 10]$ is not a closed interval so we cannot immediately conclude the extreme values of f exist. From the graph of f below (it is a portion of a parabola) we see that the extreme values do exist: the absolute maximum is 42 at $x = 6$ and the absolute minimum is 26 at $x = 10$.

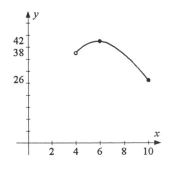

The Mean Value Theorem

Concepts to Master

Rolle's Theorem; Mean Value Theorem

Summary and Focus Questions

The Mean Value Theorem is very important because it contains an equation that relates function values and derivative values. Rolle's Theorem is a special case:

Rolle's Theorem:

If the function f
    1) is continuous on $[a, b]$
    2) is differentiable on (a, b)
    3) has $f(a) = f(b)$
then there is at least one number
$c \in (a, b)$ such that $f'(c) = 0$.

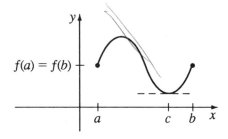

In terms of the graph of f, Rolle's Theorem says there is at least one horizontal tangent line.

Mean Value Theorem:

If the function f
    1) is continuous on $[a, b]$
    2) is differentiable on (a, b)
then there is at least one $c \in (a, b)$
such that $f'(c) = \frac{f(b) - f(a)}{b - a}$.

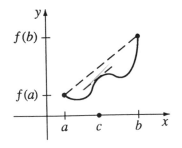

Since $\frac{f(b)-f(a)}{b-a}$ is the slope of the line joining the endpoints of the graph, the Mean Value Theorem simply says there is at least one point on the graph somewhere between the endpoints where the tangent line has the same slope as the line joining the endpoints.

If $f'(x) = g'(x)$ for all $x \in (a, b)$, then $f(x) = g(x) + c$ where c is a constant.

1) For $f(x) = 1 - x^2$ on $[-2, 1]$, do the hypotheses and conclusion of Rolle's Theorem hold?

$f'(x) = -2x$

$1 - (-2)^2$

$1 - 4 = -3 \neq 0$

f is continuous on $[-2, 1]$ and differentiable on $(-2, 1)$ but since $f(-2) = -3 \neq 0 = f(1)$ one hypothesis fails. The conclusion holds because $0 \in (-2, 1)$ and $f'(0) = 0$.

2) Do the hypotheses and conclusion of the Mean Value Theorem hold for $f(x) = 6x + x^2$ on $[1, 3]$?

The hypotheses hold because f is a polynomial and therefore continuous and differentiable everywhere. Since the hypotheses are true, the conclusion must be true as well.
$\frac{f(3)-f(1)}{3-1} = \frac{27-7}{3-1} = 10$.
$f'(x) = 6 + 2x$ which equals 10 for $c = 2$ in $(1, 3)$.

3) Mark on the x-axis the point(s) c in the
    conclusion of the Mean Value
    Theorem for the function below.

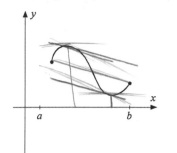

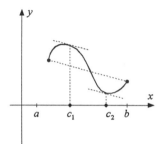

There are two points labelled c_1 and c_2 that
satisfy the conclusion of the Mean Value
Theorem.

4) David is driving on a highway which
    has a speed limit of 55 mi/hr. At 2 p.m.
    he is at milepost 110 and at 5 p.m. he
    is at milepost 290. Could David be
    proved guilty of speeding?

$$f(2)\,110 \qquad \frac{f(5)-f(2)}{5-2}=\frac{110-290}{60}$$

$$f(5)\,290 \qquad \qquad 5-2$$

$$60\,myL$$

Yes. Let $f(t)$ be his position at time t.
$f(2) = 110$ and $f(5) = 290$. The mean
value on $[2, 5]$ is $\frac{f(5)-f(2)}{5-2} = \frac{290-110}{5-2} = 60$.
By the Mean Value Theorem there is a time
$c \in (2, 5)$ such that $f'(c) = 60$. So at least
once (at time c) David's velocity was 60
mi/hr.

5) Find all numbers c that satisfy the
    conclusion of the Mean Value
    Theorem for $f(x) = x^3 - 3x^2 + x$ on
    $[0, 3]$.

The mean value on $[0, 3]$ is
$\frac{f(3)-f(0)}{3-0} = \frac{3-0}{3-0} = 1.$
$f'(x) = 3x^2 - 6x + 1.$
$3x^2 - 6x + 1 = 1$
$3x^2 - 6x = 0$
$3x(x - 2) = 0.$
Thus $f'(x) = 1$ at $x = 0$, $x = 2$. However, $0 \notin (0, 3)$ so $c = 2$ is the only point satisfying the Mean Value Theorem.

6) Suppose $f'(x) \le 2$ for all x. If $f(1) = 8$ what is the largest possible value that $f(5)$ could be?

$(1, 5)$

$f'(c) = \frac{f(5)-f(1)}{5-1} = \frac{f(5)-8}{4}$

$f'(c) \le 2$

$\frac{f(5)-8}{4} \le 2$

$f(5) - 8 \le 8$ $\boxed{f(5) \le 16}$

Since $f'(x)$ exists everywhere we can use the mean Value Theorem for f on $[1, 5]$.
For some $c \in (1, 5)$,
$f'(c) = \frac{f(5)-f(1)}{5-1} = \frac{f(5)-8}{4}$
Since $f'(c) \le 2$
$\frac{f(5)-8}{4} \le 2$
$f(5) - 8 \le 8$
$f(5) \le 16$
So $f(5)$ can be no more than 16.

7) True, False:
If $f'(x) = g'(x)$, then $f(x) = g(x)$.

nope

False. For example, $f(x) = x^2$ and $g(x) = x^2 + 6$ have the same derivative, $2x$.

Monotonic Functions and the First Derivative Test

Concepts to Master

A. Increasing, decreasing, monotonic functions; Relationship of the derivative to increasing and decreasing

B. The First Derivative Test

Summary and Focus Questions

A. A function f is <u>increasing</u> on an interval I if for all x_1, x_2, $\in I$, $x_1 < x_2$ implies $f(x_1) < f(x_2)$. In other words, as x gets larger, so does $f(x)$. f is <u>decreasing</u> means that $x_1 < x_2$ implies $f(x_1) > f(x_2)$ for all x_1, x_2, $\in I$. f is <u>monotonic</u> on I if f is either increasing on I or decreasing on I.

If f is continuous on $[a, b]$ and differentiable on (a, b):
a) $f'(x) > 0$ for all $x \in (a, b)$ implies f is increasing on $[a, b]$.
b) $f'(x) < 0$ for all $x \in (a, b)$ implies f is decreasing on $[a, b]$.

1) Given the graph of f below, on what intervals is f increasing? decreasing?

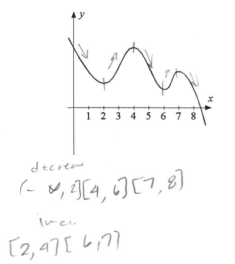

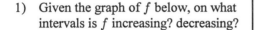

decrea
$(-\infty, 2][4, 6][7, 8)$

inc.
$[2, 4][6, 7]$

f is decreasing on each of these intervals:
$(-\infty, 2]$, $[4, 6]$, $[7, \infty)$. f is increasing on $[2, 4]$ and on $[6, 7]$.

2) Is $f(x) = x^2 + 6x$ increasing on $[-1, 2]$?

$f'(x)\ 2x+6$ $x \leq [1,2)$

$2x+6 > 0$
$\quad x > -3$ increasing

Yes. $f'(x) = 2x + 6$.
$2x + 6 > 0$ when $2x > -6$ or $x > -3$.
Thus for all $x \in [-1, 2]$, $f'(x) > 0$.
Therefore f is increasing on $[-1, 2]$.

3) Where is $f(x) = x^3 - 3x^2$ increasing and where is it decreasing?

$f'(x) = 3x^2 - 6x = 3x(x - 2) = 0$ at $x = 0$, $x = 2$. At other values of x, $f'(x)$ will be either positive or negative.

	$3x$	$x - 2$	$3x(x - 2)$
$x < 0$	$-$	$-$	$+$
$0 < x < 2$	$+$	$-$	$-$
$2 < x$	$+$	$+$	$+$

f is increasing on $(-\infty, 0)$ and $(2, \infty)$; f is decreasing on $(0, 2)$.

4) Sketch a graph of a differentiable function having all these properties: $f(0) = 1$, $f(2) = 3$, $f(5) = 0$, $f'(x) > 0$ for $x \in (0, 2)$ and $x \in (5, \infty)$, $f'(x) < 0$ for $x \in (-\infty, 0)$ and $x \in (2, 5)$.

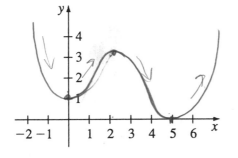

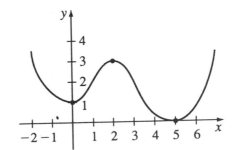

B. The First Derivative Test:

Let f be continuous on $[a, b]$ and differentiable on (a, b). Let c be a critical number for f in (a, b).

    a) If $f'(x)$ is positive in (a, c) and negative in (c, b), then f has a local maximum at c.

    b) If $f'(x)$ is negative in (a, c) and positive in (c, b), f has a local minimum at c.

    c) If $f'(x)$ has the same sign (either positive or negative) in both (a, c) and (c, b), $f(c)$ is not a local extrema.

5) Suppose 8 is a critical number for a function f with $f'(x) > 0$ for $x \in (8, 9)$ and $f'(x) < 0$ for $x \in (7, 8)$. Then 8 is:

    a) a local maximum
    b) a local minimum
    c) not a local extremum

b), by the First Derivative Test.

6) What can you conclude about the local extrema of a differential function f from the following? Critical numbers for f are $x = 1, 3, 4, 6, 8$.

Interval	Sign of $f'(x)$
$(-\infty, 1)$	positive
$(1, 3)$	negative
$(3, 4)$	negative
$(4, 6)$	positive
$(6, 8)$	positive
$(8, \infty)$	negative

f has a local maximum at $x = 1$ and $x = 8$; f has a local minimum at $x = 4$; there are no local extrema at $x = 3$ or $x = 6$.

7) Is $x = 5$ a local minimum for $f(x) = 10x - x^2 + 1$?

No. $f'(x) = 10 - 2x > 0$ for all $x < 5$ while $f'(x) < 0$ for all $x > 5$. f has a local maximum at $x = 5$.

8) Find where the local maximum and minimum values of $f(x) = 4x^3 - x^4$ occur.

$f'(x) = 12x^2 - 4x^3 = 4x^2(3 - x)$
$f'(x) = 0$ at $x = 0$, $x = 3$, so there are two critical numbers. Since $f'(x) \geq 0$ for $x > 3$ and $f'(x) \leq 0$ for $x > 3$, 3 is a local maximum. No extreme value occurs at $x = 0$.

9) Find where the local maximum and minimum values of $f(x) = x + \sin x$ occur.

$f'(x) = 1 + \cos x$. Since $-1 \leq \cos x \leq 1$, $f'(x) \geq 0$ for all x. Therefore, f has no local extreme values.

Concavity and Points of Inflection

Concepts to Master

A. Concave upward and downward; Test for concavity
B. Point of inflection; The Second Derivative Test

Summary and Focus Questions

A. A function f is <u>concave upward</u> on an interval I if f lies above all tangent lines to f in I. f is <u>concave downward</u> on I if the graph of f is below all tangent lines.

concave upward   concave downward

A test for concavity involves the second derivative:

If f is continuous on $[a, b]$ and twice differentiable on (a, b) (meaning $f''(x)$ exists for all $x \in (a, b)$) then:
    a) If $f''(x) > 0$ for all $x \in (a, b)$, f is concave *upward* on $[a, b]$.
    b) If $f''(x) < 0$ for all $x \in (a, b)$, f is concave *downward* on $[a, b]$.

1) Use the graph below to answer true or false to each.

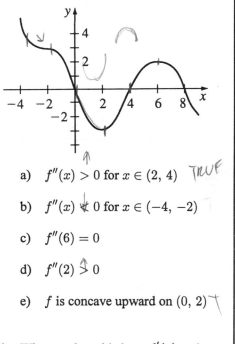

a) $f''(x) > 0$ for $x \in (2, 4)$ ᴛʀᴜᴇ

True.

b) $f''(x) \not< 0$ for $x \in (-4, -2)$

False.

c) $f''(6) = 0$

False. $f''(6)$ is negative.

d) $f''(2) \gtrless 0$

True.

e) f is concave upward on $(0, 2)$

True.

2) What can be said about $f'(x)$ and $f''(x)$ for each.

a)

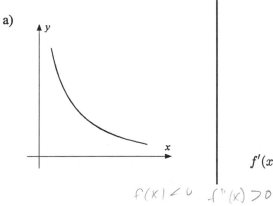

$f'(x) < 0$ and $f''(x) > 0$

$f(x) < 0$ $f''(x) > 0$

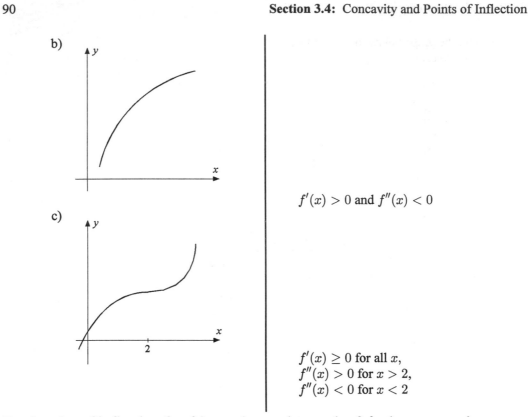

b)

c)

$f'(x) > 0$ and $f''(x) < 0$

$f'(x) \geq 0$ for all x,
$f''(x) > 0$ for $x > 2$,
$f''(x) < 0$ for $x < 2$

B. A <u>point of inflection</u> for f is a point on the graph of f where concavity changes from concave downward to concave upward or from concave upward to concave downward.

Concave downward to concave upward

Concave upward to concave downward

The Second Derivative Test is:

If f'' is continuous on (a, b) containing c and $f'(c) = 0$, then
    a) if $f''(c) < 0$, f has a local maximum at c.
    b) if $f''(c) > 0$, f has a local minimum at c.

If it should happen that both $f'(c) = 0$ and $f''(c) = 0$ then no conclusion can be made about an extremum at c.

3) What are the points of inflection for the function in question 1)?

                $x = -2, 0, 4, 8$

4) Find the points of inflection for $f(x) = 8x^3 - x^4$.

      $f'(x) = 24x^2 - 4x^3$,
      $f''(x) = 48x - 12x^2 = 12x(4 - x) = 0$
      at $x = 0$, $x = 4$. Since f is a polynomial the only points of inflection are at 0 and 4.

5) Find all local extrema of
    a) $f(x) = x^3 - 6x^2 - 36x$.

      $f'(x) = 3x^2 + 12x - 36$
           $= 3(x - 2)(x + 6)$.
      The critical points are at $x = 2, -6$.
      $f''(x) = 6x + 12$.
      At $x = 2$, $f''(2) = 24 > 0$ so f has a local minimum at 2.
      At $x = -6$, $f''(-6) = -24 < 0$ so f has a local maximum at -6.

b) $f(x) = x^2 + 2\cos x$.

$f'(x) = 2x - 2\sin x = 0$ when $x = \sin x$. The only solution to this equation is $x = 0$. $f''(x) = 2 - 2\cos x$ but $f''(0) = 0$ so the Second Derivative Test fails to give additional information. However $f''(x) \geq 0$, for all x, means f is concave upward and thus f has a local minimum at 0.

6) Sketch a graph of a function f having all these properties:
$f(-1) = 4$, $f(0) = 2$, $f(2) = 1$, $f(3) = 0$
$f'(x) \leq 0$ for $x < 3$ and
$f'(x) \geq 0$ for $x > 3$.
$f''(x) < 0$ for $0 < x < 2$ and
$f''(x) \geq 0$ elsewhere.

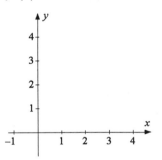

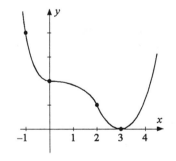

Section 3.5

Limits at Infinity; Horizontal Asymptotes

Concepts to Master

A. Define and evaluate limits at infinity
B. Horizontal asymptotes

Summary and Focus Questions

A. $\lim_{x \to \infty} f(x) = L$ means that as x is assigned larger values without bound, the corresponding values of $f(x)$ approach L. The formal definition of $\lim_{x \to \infty} f(x) = L$ is that for each $\varepsilon > 0$ there exists a number N such that $|f(x) - L| < \varepsilon$ whenever $x > N$. $\lim_{x \to \infty} f(x) = L$ describes what happens to the graph of f out on the "far right side": the graph approaches the line $y = L$.

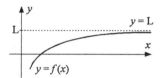

$\lim_{x \to -\infty} f(x) = L$ is defined similarly and means that $f(x)$ approaches L as x takes more negative values without bound.

Most of the techniques for evaluating ordinary limits may be used for limits at infinity. In addition, for a limit involving a rational function, dividing the numerator and denominator by the highest power of x that appears in the denominator may help, as in:

$$\lim_{x \to \infty} \frac{4x^2 + 7x + 5}{2x^2 + 3} = \lim_{x \to \infty} \frac{\frac{(4x^2 + 7x + 5)}{x^2}}{\frac{(2x^2 + 3)}{x^2}} = \lim_{x \to \infty} \frac{4 + \frac{7}{x} + \frac{5}{x^2}}{2 + \frac{3}{x^2}} = \frac{4 + 0 + 0}{2 + 0} = 2$$

All of the usual limit theorems hold for limits at infinity.

Similar to the infinite limits in section 1.4, $\lim\limits_{x\to\infty} f(x) = \infty$ means that as x increases without bound, then so do the corresponding $f(x)$ values.

1) $\lim\limits_{x\to\infty} f(x) = L$ means for all _____
   there exists _____ such that _____
   whenever _____.

$\varepsilon > 0$
N
$|f(x) - L| < \varepsilon$
$x > N$

2) Evaluate

a) $\lim\limits_{x\to\infty} \dfrac{5x^3+7x+1}{3x^3+2x^2+3}$

Divide both numerator and denominator by x^3.
$$\lim\limits_{x\to\infty} \frac{5+\frac{7}{x^2}+\frac{1}{x^3}}{3+\frac{2}{x}+\frac{3}{x^3}} = \frac{5+0+0}{3+0+0} = \frac{5}{3}.$$
Limits of this type can also be "eye-balled:"
by observing that for very large positive x,
$5x^3 + 7x + 1 = 1 \approx 5x^3$ and
$3x^3 + 2x^2 + 3 \approx 3x^3$. Thus the limit is
$\lim\limits_{x\to\infty} \dfrac{5x^3}{3x^3} = \dfrac{5}{3}.$

b) $\lim\limits_{x\to-\infty} \dfrac{2x+5}{\sqrt{x^2+4}}$

$$\lim\limits_{x\to-\infty} \frac{\frac{2x+5}{(-x)}}{\frac{\sqrt{x^2+4}}{(-x)}} = \lim\limits_{x\to-\infty} \frac{\frac{2x+5}{(-x)}}{\frac{\sqrt{x^2+4}}{\sqrt{x^2}}}$$
(Remember, since $x \to -\infty$, $x < 0$, so $\sqrt{x^2} = -x$.)
$$= \lim\limits_{x\to-\infty} \frac{-2-\frac{5}{x}}{\sqrt{1+\frac{4}{x^2}}} = \frac{-2-0}{\sqrt{1+0}} = -2.$$

c) $\lim\limits_{x\to\infty} \dfrac{\cos x}{x}$

Since $-1 \le \cos x \le 1$, $\dfrac{-1}{x} \le \dfrac{\cos x}{x} \le \dfrac{1}{x}$.
Both $\lim\limits_{x\to\infty} \dfrac{-1}{x} = 0$ and $\lim\limits_{x\to\infty} \dfrac{1}{x} = 0$.
Therefore $\lim\limits_{x\to\infty} \dfrac{\cos x}{x} = 0$.

d) $\lim\limits_{x\to\infty} \dfrac{x^3+6x+1}{2x^2-5x}$

Dividing the numerator and denominator by x^2, we have $\lim\limits_{x\to\infty} \dfrac{x+\frac{6}{x}+\frac{1}{x^2}}{2-\frac{5}{x}}$.

As x grows larger, the denominator approaches 2 but the numerator grows larger, so $\lim\limits_{x\to\infty} \dfrac{x^3+6x+1}{2x^2-5x} = \infty$.

3) True or False:
If all three limits exist, then
$$\lim_{x\to\infty} (f(x)g(x))$$
$$= \left(\lim_{x\to\infty} f(x) \right) \left(\lim_{x\to\infty} g(x) \right)$$

True.

B. A <u>horizontal asymptote</u> for a function f is a line $y = L$ such that $\lim\limits_{x\to\infty} f(x) = L$, $\lim\limits_{x\to-\infty} f(x) = L$, or both. A function can have at most two horizontal asymptotes. Here are the graphs of three examples:

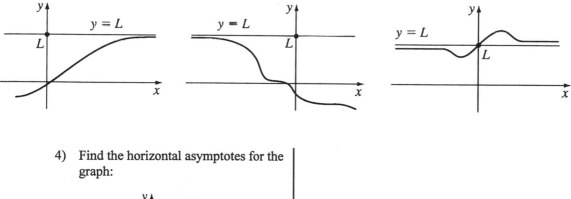

4) Find the horizontal asymptotes for the graph:

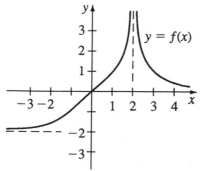

f has two horizontal asymptotes, $y = -2$ and $y = 0$. (Note: $x = 2$ is a vertical asymptote.)

5) Find the horizontal asymptotes for
$f(x) = \frac{x}{|x|-1}$.

$y = 1$, because $\lim\limits_{x \to \infty} \frac{x}{|x|-1} = 1$ and $y = -1$, because $\lim\limits_{x \to -\infty} \frac{x}{|x|-1} = -1$.

6) Find the asymptotes and sketch the graph of $f(x) = \frac{x}{x^2-1}$.

$\lim\limits_{x \to \infty} \frac{x}{x^2-1} = 0$ and $\lim\limits_{x \to -\infty} \frac{x}{x^2-1} = 0$, so $y = 0$ is the only horizontal asymptote. $\lim\limits_{x \to 1^+} \frac{x}{x^2-1} = \infty$ and $\lim\limits_{x \to 1^-} \frac{x}{x^2-1} = -\infty$ so $x = 1$ is a vertical asymptote. Likewise $x = -1$ is a vertical asymptote because $\lim\limits_{x \to -1^+} \frac{x}{x^2-1} = \infty$ and $\lim\limits_{x \to -1^-} \frac{x}{x^2-1} = -\infty$.
$\frac{x}{x^2-1} = 0$ only at $x = 0$, so $x = 0$ is the only intercept.
$f'(x) = \frac{(x^2-1)-x(2x)}{(x^2-1)^2} = -\frac{x^2+1}{(x^2-1)^2}$ which is always negative when it exists. Thus $f(x)$ is always decreasing.
$f''(x) = -\frac{(x^2-1)^2 2x - (x^2+1)(2(x^2-1)2x)}{(x^2-2)^4}$
$\quad\quad = \frac{2x(x^3-x^2-x+1)}{(x^2-1)^3}$
$f''(x) = 0$ at $x = 0$. 0 is a point of inflection. Finally, the graph is:

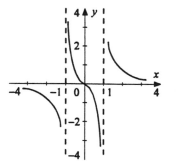

7) Find the vertical and horizontal asymptotes of $f(x) = \frac{1-\cos x}{x}$.

$f(x)$ is not defined for $x = 0$, so $x = 0$ could be a vertical asymptote. However, $\lim\limits_{x \to 0} \frac{1-\cos x}{x} = 0$, so f has no vertical asymptotes. To find limits at infinity we use the Squeeze Theorem. $-1 \le \cos x \le 1$. Thus, $-1 \le -\cos x \le 1$ and $0 \le 1 - \cos x \le 2$.

Thus, $\frac{0}{x} \leq \frac{1-\cos x}{x} \leq \frac{2}{x}$, so
$0 \leq \frac{1-\cos x}{x} \leq \frac{2}{x}$ for $x > 0$.
Since $\lim\limits_{x \to \infty} \frac{2}{x} = 0$, $\lim\limits_{x \to \infty} \frac{1-\cos x}{x} = 0$.
In a similar manner we can see that
$\lim\limits_{x \to -\infty} \frac{1-\cos x}{x} = 0$. Thus, $y = 0$ is the only
horizontal asymptote.

Curve Sketching

Concepts to Master

A. Curve sketching using information obtained through calculus concepts
B. Slant asymptotes

Summary and Focus Questions

A. You should develop a checklist of features to consider before drawing a graph. Questions to answer in sketching a graph of $y = f(x)$ include:

A. Domain of f For what x is $f(x)$ defined?

B. x-intercept(s)   What are the solution(s) (if any) to $f(x) = 0$?
    y-intercept   What value (if any) is $f(0)$?

C. Symmetry about y-axis   Is $f(-x) = f(x)$?
    Symmetry about origin   Is $f(-x) = -f(x)$?
    Periodic   Is there a number p such that $f(x + p) = f(x)$ for all x in the domain?

D. Horizontal asymptote(s)   Does $\lim\limits_{x \to \infty} f(x)$ or $\lim\limits_{x \to -\infty} f(x)$ exist?
    Vertical asymptote(s)   For what a is $\lim\limits_{x \to a^+} f(x) = \infty$ or $-\infty$?
        For what a is $\lim\limits_{x \to a^-} f(x) = \infty$ or $-\infty$?

E. Increasing   On what intervals is $f'(x) \geq 0$?
    Decreasing   On what intervals is $f'(x) \leq 0$?

F. Critical numbers   Where does $f'(x) = 0$ or not exist?
    Local extrema   Where are the local maxima or minima (if any)? (Use the First and Second Derivative Tests.)

G. Concave upward   On what intervals is $f''(x) \geq 0$?
    Concave downward   On what intervals is $f''(x) \leq 0$?
    Inflection points   Where does f change concavity?
                                Where is $f''(x) = 0$?

1) Sketch a graph of $y = f(x)$ that has all
these properties:
     domain $= (-\infty, -1) \cup (1, \infty)$
     $f(2) = 0$
     $f(x) = -f(-x)$
     $\lim\limits_{x \to \infty} f(x) = 4$ and $\lim\limits_{x \to 1^+} f(x) = -\infty$
     $f'(3) = 0$, $f'(5) = 0$
     $f'(x) \geq 0$ on $(1, 3) \cup (5, \infty)$
     $f'(x) \leq 0$ on $(3, 5)$
     local maximum at 3
     local minimum at 5
     $f''(x) \geq 0$ on $(4, 6)$
     $f''(x) \leq 0$ on $(1, 4) \cup (6, \infty)$
     $f''(4) = 0$, $f''(6) = 0$

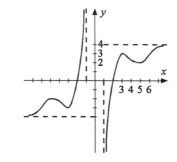

2) Sketch a graph of $f(x) = \frac{1}{x^2-4}$.

A. The domain is all x except where
$x^2 - 4 = 0$; all x except $x = 2$,
$x = -2$.

B. $\frac{1}{x^2-4}$ is never 0, so no x-intercepts.
At $x = 0$, $y = \frac{1}{0^2-4} = \frac{-1}{4}$,
the y-intercept.

C. $f(-x) = \frac{1}{(-x)^2-4} = \frac{1}{x^2-4} = f(x)$, so
there is symmetry about the y-axis.
But $f(-x) \neq -f(x)$, so no symmetry
about $(0, 0)$.

D. $\lim\limits_{x\to\infty} \frac{1}{x^2-4} = 0$ and $\lim\limits_{x\to-\infty} \frac{1}{x^2-4} = 0$,
$y = 0$ is the only horizontal
asymptote. $\lim\limits_{x\to 2^+} \frac{1}{x^2-4} = \infty$ and
$\lim\limits_{x\to 2^-} \frac{1}{x^2-4} = \infty$, so $x = 2$ and $x = -2$
are vertical asymptotes.

E. $f'(x) = (-1)(x^2 - 4)^{-2}(2x)$
$= -2x(x^2 - 4)^{-2} = \frac{-2x}{(x^2-4)^2}$.
For $x > 0$, $x \neq 2$, $f'(x) < 0$ so f is
decreasing on $(0, 2)$ and $(2, \infty)$.
For $x < 0$, $x \neq -2$, $f'(x) > 0$ so f is
increasing on $(-\infty, -2)$ and $(-2, 0)$.

F. $f'(x) = \frac{-2x}{(x^2-4)^2} = 0$ at $x = 0$.
$f''(x) = (-2x)[-2(x^2 - 4)^{-3} \cdot 2x]$
$\quad + (x^2 - 4)^{-2}(-2)$
$= (x^2 - 4)^{-3}[8x^2 - 2(x^2 - 4)]$
$= \frac{6x^2+4}{(x^2-4)^3}$.
$f''(0) = \frac{4}{(-4)^3} = -\frac{1}{16}$. Thus f has a
local maximum at $x = 0$.

G. Since $f''(x) = \frac{6x^2+4}{(x^2-4)^3}$, the sign of
$f''(x)$ is determined by $(x^2 - 4)^3$ and
thus by $x^2 - 4$.
$x^2 - 4 > 0$ for $x > 2$ and $x < -2$
and $x^2 - 4 < 0$ for $-2 < x < 2$.
f is concave upward on $(-\infty, -2)$
and $(2, \infty)$ and concave downward on
$(-2, 2)$.

H. Finally, a sketch:

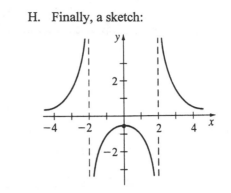

B. $y = mx + b$ is a <u>slant asymptote</u> of $y = f(x)$ means that
$$\lim_{x \to \infty} [f(x) - (mx + b)] = 0 \text{ or } \lim_{x \to -\infty} [f(x) - (mx + b)] = 0.$$

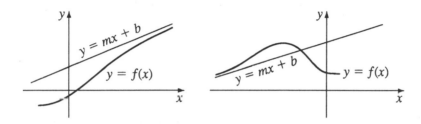

A slant asymptote is a nonvertical, nonhorizontal line that the curve approaches for large (positive or negative) values of x.

3) Find a slant asymptote for
$f(x) = \frac{x^2+x+1}{x}$.

$f(x) = \frac{x^2+x+1}{x} = x + 1 + \frac{1}{x}$.
The line $y = x + 1$ is a slant asymptote.

4) Can a polynomial of degree greater than one have a slant asymptote?

No. If $f(x)$ is a polynomial whose degree is two or more then $f(x) - (mx + b)$ will still have degree two or more. Thus, $\lim_{x \to \infty} [f(x) - (mx + b)]$ will not be zero (it will be ∞ or $-\infty$).

Graphing with Calculus and Calculators

Concepts to Master

A. Use calculus to improve information from graphing calculator displays
B. Determine the graphs of families of functions

Summary and Focus Questions

A. A graphing calculator can display an initial graph of a function. The concepts of calculus can then be used to refine it so that details such as relative extrema and concavity stand out.

1) Use a graphing calculator to graph
$f(x) = 9x^4 - 76x^3 + 180x^2$. Then use calculus to refine it.

An initial graph from a calculator may look something like:

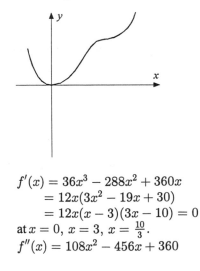

$f'(x) = 36x^3 - 288x^2 + 360x$
$\qquad = 12x(3x^2 - 19x + 30)$
$\qquad = 12x(x - 3)(3x - 10) = 0$
at $x = 0$, $x = 3$, $x = \frac{10}{3}$.
$f''(x) = 108x^2 - 456x + 360$

x	f'	f''	Type
0	0	360	relative minimum
3	0	-36	relative maximum
$\frac{10}{3}$	0	36	relative minimum

The graph of f with its critical numbers 0, 3, and $\frac{10}{3}$ is:

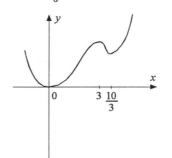

B. A <u>family</u> of functions is a collection of functions defined by a formula with one or more arbitrary constants. Varying the values of the constants results in changes in the graphs of the functions.

For example, the family described by $f(x) = x^2 + bx + 1$ has graphs which are parabolas passing through $(0,\ 1)$ with vertex (relative minimum) at $x = -\frac{b}{2}$.

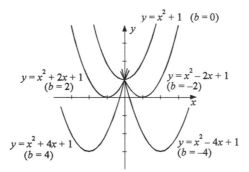

2) Describe the graphs of the family of functions $f(x) = x^3 - 3ax$.

Each function passes through $(0,\ 0)$. $f'(x) = 3(x^2 - a)$. For $a > 0$, f has a relative maximum at $x = -\sqrt{a}$ and relative minimum at $x = \sqrt{a}$. For $a < 0$, f has no extrema.

$f''(x) = 6x$ so f is concave up for $x > 0$ and concave down for $x < 0$. Thus $x = 0$ is an inflection point. The family of graphs is:

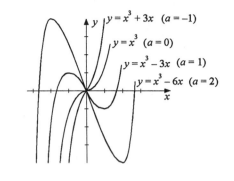

Applied Maximum and Minimum Problems

Concepts to Master

Solutions to applied extremum problems

Summary and Focus Questions

Many problems can be characterized as "find the maximum (or minimum) value of some quantity Q subject to some given conditions." For example, suppose we wish to find the dimensions of the largest rectangular field enclosed by 100 feet of fencing. The quantity to be maximized is the area, subject to the condition that the perimeter is 100.

A procedure to help solve problems of this type is:
  (1) Introduce notation identifying the quantity to be maximized (or minimized), Q, and other variable(s). Here a diagram may help.
  (2) Write Q as a function of the other variables. Also, express any relationships among the other variables with equations.
  (3) Rewrite, if necessary, Q as a function of just one of the variables, using given relationships and determine the domain D of Q.
  (4) Find the absolute extrema of Q on domain D.

In our example above, the steps in its solution are:
  (1) Let $A = $ area, $x = $ width,
      $y = $ height
  (2) $A = xy$, $2x + 2y = 100$
  (3) From $2x + 2y = 100$,
      $y = 50 - x$. Thus,
      $A = x(50 - x) = 50x - x^2$.
      The domain is $x \in [0, 50]$.

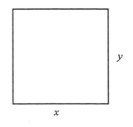

(4) $A'(x) = 50 - 2x = 0$ at $x = 25$. Thus, $y = 50 - 25 = 25$. So the largest area enclosed by 100 feet of fencing is a square 25 ft by 25 ft.

Steps 2 and 3 in the procedure are often the most difficult because you may need to recall facts from geometry, trigonometry, etc.

1) A gardener wishes to enclose a rectangular area with 300 feet of fence and fence it down the middle to divide it into two equal subareas. What is the largest rectangular area that may be enclosed?

(1) Let A be the area of the enclosed rectangle.

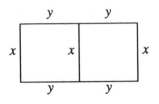

Label one side as x and the half side of the other as y (since that side is bisected).

(2) $A = x(2y) = 2xy$.
The total amount of fence is $3x + 4y = 300$. Thus, $y = \frac{300 - 3x}{4}$.

(3) Therefore,
$A = 2xy = 2x\frac{300 - 3x}{4} = 150x - \frac{3}{2}x^2$.
Since $3x + 4y = 300$ and both $x \geq 0$ and $y \geq 0$, $x \in [0, 100]$.

(4) Maximize $A = 150x - \frac{3}{2}x^2$ on $[0, 100]$. Clearly 0 and 100 produce minimum values, so A is maximum when $A' = 0$. $A' = 150 - 3x = 0$ at $x = 50$. Then $y = \frac{300 - 3(50)}{4} = 37.5$. Thus the maximum value of A is $2xy = 2(50)(37.5) = 3750$ ft^2.

2) A cylindrical aluminum cup is to be made from 12 square inches of pressed aluminum. What is the largest possible volume of such a cup?

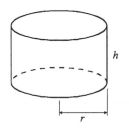

(1) Maximize the volume V of a cup whose circular base has radius r and whose height is h.

(2) $V = \pi r^2 h$. The total surface area (bottom plus sides) is $12 = \pi r^2 + 2\pi rh$. Thus $h = \frac{12 - \pi r^2}{2\pi r}$.

(3) Thus $V = \pi r^2 \frac{12 - \pi r^2}{2\pi r}$. $V = 6r - \frac{\pi}{2}r^3$, where $r \in \left[0, \sqrt{\frac{12}{\pi}}\right]$.

(4) $V' = 6 - \frac{3\pi}{2}r^2 = 0$

$r^2 = \frac{4}{\pi}$

$r = \frac{2}{\sqrt{\pi}}$

$V'' = -3\pi r$ so $r = \frac{2}{\sqrt{\pi}}$ is a maximum.

At , $r = \frac{2}{\sqrt{\pi}}$ the maximum volume is

$V = 6\left(\frac{2}{\sqrt{\pi}}\right) - \frac{\pi}{2}\left(\frac{2}{\sqrt{\pi}}\right)^3 = \frac{8}{\sqrt{\pi}}$ in^3.

Antiderivatives

Concepts to Master

General and particular antiderivatives of a function

Summary and Focus Questions

$F(x)$ is an <u>antiderivative</u> of the function $f(x)$ means $F'(x) = f(x)$. For example, $F(x) = 3x^4$ is an antiderivative of $f(x) = 12x^3$ because $(3x^4)' = 12x^3$.

If F is an antiderivative of f then <u>all</u> other antiderivatives of f have the form $F(x) + C$, where C is a constant.

Here are some antidifferentiation formulas that you should know well:

Function	Form of All Antiderivatives
x^n (except $n = -1$)	$\frac{x^{n+1}}{n+1} + C$
$\sin x$	$-\cos x + C$
$\cos x$	$\sin x + C$
$\sec^2 x$	$\tan x + C$
$\sec x \tan x$	$\sec x + C$

The antiderivative of $f \pm g$ is the antiderivative of f plus or minus the antiderivative of g, and the antiderivative of $cf(x)$ is c times the antiderivative of $f(x)$ (c, any constant).

The general solution to the differential equation $\frac{dy}{dx} = f(x)$ is all antiderivatives of f. To find a particular solution, first find the general solution, then substitute the given values to determine the constant C.

Example:

The general solution to $\frac{dy}{dx} = 10x$ is $y = 5x^2 + C$. If we are also given that $y = 4$ when $x = 2$, then substituting we have $4 = 5(2)^2 + C$, so $C = -16$. The particular solution is $y = 5x^2 - 16$.

1) If $g(x)$ is an antiderivative of $h(x)$,
    then ____'$(x) =$ ____(x).

$g'(x) = h(x)$

2) Find all antiderivatives of:
    a) $f(x) = x^7$

$\frac{x^8}{8} + C$

    b) $f(x) = \cos x - \sec^2 x$

$\sin x - \tan x + C$

    c) $f(x) = x + x^{-2}$

$\frac{x^2}{2} - \frac{1}{x} + C$

3) Find $f(x)$ where $f(2) = 3$ and
    $f'(x) = 4x + 5$.

$f(x) = 2x^2 + 5x + C$
$f(2) = 2(2)^2 + 5(2) + C = 18 + C$
$18 + C = 3$ so $C = -15$
$f(x) = 2x^2 + 5x - 15$

4) A particle moves along a scale with
    velocity $v = 3t + 7$. If the particle is at
    4 on the scale at time $t = 1$, find the
    position function $s(t)$.

$s(t)$ is an antiderivative of $v(t) = 3t + 7$.
$s(t) = \frac{3}{2}t^2 + 7t + C$
$4 = \frac{3}{2}(1)^2 + 7(1) + C$
$C = -\frac{9}{2}$
$s(t) = \frac{3}{2}t^2 + 7t - \frac{9}{2}$

4

Integrals

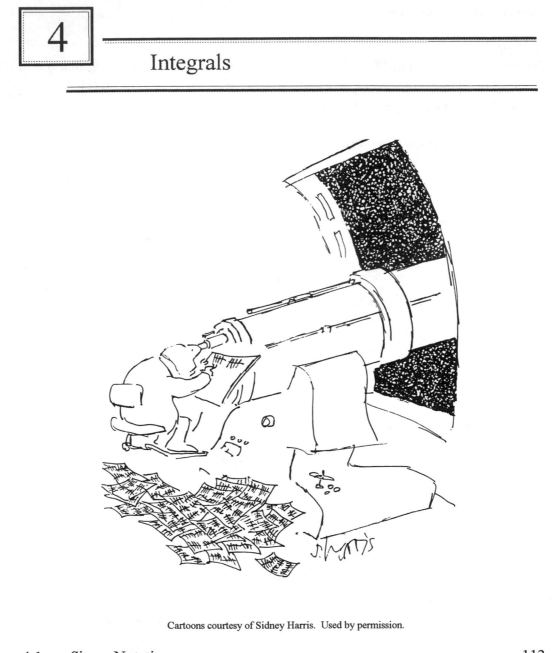

Cartoons courtesy of Sidney Harris. Used by permission.

Sigma Notation

Concepts to Master

Evaluation of sums using sigma notation

Summary and Focus Questions

For real numbers a_m, a_{m+1}, ... a_n,

$$\sum_{i=m}^{n} a_i = a_m + a_{m+1} + \ldots + a_n.$$

This is a compact notation for the sum of the numbers a_m through a_n.

$\sum_{i=m}^{n} f(i)$ is the number found by determining the numbers $f(m)$, $f(m+1)$, ..., $f(n)$ and then adding them up, as in:

$$\sum_{i=2}^{5} i^2 = (2)^2 + (3)^2 + (4)^2 + (5)^2 = 4 + 9 + 16 + 25 = 54.$$

Some rules for manipulating sigma notation sums are:

1. $\displaystyle\sum_{i=m}^{n} c\,a_i = c\left(\sum_{i=m}^{n} a_i\right)$, where c is a constant.

2. $\displaystyle\sum_{i=m}^{n} (a_i \pm b_i) = \sum_{i=m}^{n} a_i \pm \sum_{i=m}^{n} b_i.$

Some summations may be rewritten as a function of their upper limit thus replacing the summation notation with a simple expression. Here are five such expressions:

$$\sum_{i=1}^{n} 1 = n \qquad\qquad \sum_{i=1}^{n} i = \frac{n(n+1)}{2}$$

$$\sum_{i=1}^{n} i^2 = \frac{n(n+1)(2n+1)}{6} \qquad\qquad \sum_{i=1}^{n} i^3 = \left(\frac{n(n+1)}{2}\right)^2$$

$$\sum_{i=1}^{n} i^4 = \frac{n(n+1)(2n+1)(3n^2+3n-1)}{30}$$

These may be combined to simplify other sums as in:

$$\sum_{i=1}^{n}(i^2 + 4i) = \sum_{i=1}^{n} i^2 + \sum_{i=1}^{n} 4i = \sum_{i=1}^{n} i^2 + 4\sum_{i=1}^{n} i$$

$$= \frac{n(n+1)(2n+1)}{6} + 4\frac{n(n+1)}{2} = \frac{n(n+1)(2n+13)}{6}$$

1) Write the following in sigma notation.

a) $2^2 + 3^2 + 4^2 + 5^2 + 6^2$

$$\sum_{i=2}^{6} i^2$$

b) $\frac{1}{2} - \frac{1}{4} + \frac{1}{8} - \frac{1}{16} + \frac{1}{32}$

$$\sum_{i=1}^{5} \frac{(-1)^{i+1}}{2^i} \left(\text{or } -\sum_{i=1}^{5}\left(-\frac{1}{2}\right)^i \right)$$

2) What number is $\displaystyle\sum_{i=1}^{4} 3i$?

From the definition, this is
$3(1) + 3(2) + 3(3) + 3(4)$
$\qquad = 3 + 6 + 9 + 12 = 30.$
It may also be evaluated as
$$\sum_{i=1}^{4} 3i = 3\sum_{i=1}^{4} i = 3\left[\frac{4(4+1)}{2}\right] = 30.$$

3) Write the following as an expression in n:
$$\sum_{i=1}^{n}(12i^2 - 4ni)$$

$$\sum_{i=1}^{n}(12i^2 - 4ni) = \sum_{i=1}^{n} 12i^2 - \sum_{i=1}^{n} 4ni$$
$$= 12\sum_{i=1}^{n} i^2 - 4n\sum_{i=1}^{n} i$$
$$= 12\frac{n(n+1)(2n+1)}{6} - (4n)\frac{n(n+1)}{2}$$
$$= n(n+1)(2n+2) = 2n(n+1)^2.$$

4) Evaluate $\displaystyle\sum_{i=1}^{10}(f(i) - f(i-1))$, where $f(i) = 2i^2 + 1$ for $i = 0, 1, 2\ldots, 10$.

We note that this is a "collapsing sum".
$(f(1) - f(0)) + (f(2) - f(1)) + (f(3)$
$\quad - f(2)) + \ldots + (f(10) - f(9)).$
All terms cancel except $-f(0)$ and $f(10)$.
The sum is
$f(10) - f(0) = 101 - 1 = 100.$

Area

Concepts to Master

Interval partition; Norm of a partition; Approximation of area under a curve; Area under a curve

Summary and Focus Questions

The goal of this section is to determine a way to find the area under a curve $y = f(x)$, above the x-axis, and between lines $x = a$ and $x = b$.

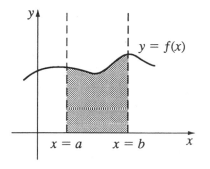

A <u>partition</u> P of $[a, b]$ is a set of subintervals $[x_0, x_1], [x_1, x_2], \ldots, [x_{n-1}, x_n]$, where $a = x_0 < x_1 < x_2 \ldots < x_{n-1} < x_n = b$. The width of the ith subinterval is $\Delta x_i = x_i - x_{i-1}$. The <u>norm of P</u>, denoted $\|P\|$, is the largest of all Δx_i.

For example, P: $[2, 4], [4, 5], [5, 8], [8, 9]$ is a partition of $[2, 9]$ and has norm 3 (the width of $[5, 8]$).

A partition of $[a, b]$ into n subintervals of equal widths has norm $\frac{b-a}{n}$.

To approximate the area described above:
    (1) Partition $[a, b]$ into n subintervals.
    (2) Select a point x_i^* in each subinterval.
    (3) Compute $\sum_{i=1}^{n} f(x_i^*) \cdot \Delta x_i$.

115

For some functions the area may be determined exactly by:

    (1) Partition $[a, b]$ into subintervals of equal length $\Delta x = \frac{b-a}{n}$.

    (2) Select a point x_i^* in each subinterval and write each x_i^* in terms of n and i. Do so consistently. For example, choosing x_i^* to be the left endpoint of $[x_{i-1}, x_i]$ means $x_i^* = a + (i - 1)\Delta x$.

    (3) Write the approximating sum $\sum\limits_{i=1}^{n} f(x_i^*) \cdot \Delta x$ as a function of n, as was done in Section 4.1.

    (4) Determine the limit of the expression found in step (3) as $n \to \infty$.

1) For the interval $[3, 8]$ the norm of the partition $[3, 4]$, $[4, 6]$, $[6, 7]$, $[7, 8]$ is

    _____.

          2 (the length of the widest subinterval).

2) True or False:
The area of a region bounded by
$y = f(x)(\geq 0)$, $y = 0$, $x = a$, $x = b$
is the limit of the sums of the areas of
approximating rectangles.

          True.

3) Using x_i^* as the midpoint of each
interval of the partition $[1, 3]$, $[3, 6]$,
$[6, 10]$ approximate the area under
$f(x) = 1 + 4x^2$ between $x = 1$ and
$x = 10$.

          Here $n = 3$.

i	subint	Δx_i	x_i^*	$f(x_i^*)$
1	$[1, 3]$	2	2	17
2	$[3, 6]$	3	4.5	82
3	$[6, 10]$	4	8	257

$$\sum_{i=1}^{3} f(x_i^*)\Delta x_i = 2(17) + 3(82) + 4(257)$$

          $= 1308$. (We shall later see that the exact area is 1341.)

4) For the area under $f(x) = 10x - x^2$ between $x = 2$ and $x = 9$ we will use the partition [2, 4], [4, 6], [6, 9] and choose $x_1^* = 3$, $x_2^* = 5$, and $x_3^* = 7$. Sketch the area and the corresponding approximating rectangles.

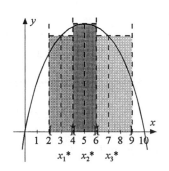

5) Find the area under $y = 10 - 2x$ between $x = 1$ and $x = 4$. Take x_i^* to be the right endpoint. Use subintervals of equal width.

With n subintervals we have
$\Delta x = \frac{b-a}{n} = \frac{4-1}{n} = \frac{3}{n}$.

$x_1 = 1 + \Delta x$, $x_2 = 1 + 2\Delta x$, and in general $x_i = 1 + i\,\Delta x$. Selecting x_i^* to be the right endpoint means $x_i^* = x_i$. Thus $x_i^* = 1 + i\,\Delta x = 1 + i\left(\frac{3}{n}\right) = 1 + \frac{3i}{n}$.
Hence $f(x_i^*) = 10 - 2\left(1 + \frac{3i}{n}\right) = 8 - \frac{6i}{n}$.
The approximating sum is
$$\sum_{i=1}^{n} f(x_i^*)\Delta x = \sum_{i=1}^{n}\left(8 - \frac{6i}{n}\right)\left(\frac{3}{n}\right) = \sum_{i=1}^{n}\left(\frac{24}{n} - \frac{18i}{n^2}\right)$$
$$= \sum_{i=1}^{n}\frac{24}{n} - \sum_{i=1}^{n}\frac{18i}{n^2} = \frac{24}{n}\sum_{i=1}^{n}1 - \frac{18i}{n^2}\sum_{i=1}^{n}i$$
$$= \frac{24}{n}(n) - \frac{18i}{n^2}\left(\frac{n(n+1)}{2}\right)$$
$$= 24 - 9\frac{n(n+1)}{n^2}.$$
Finally,
$$\lim_{n\to\infty}\left(24 - 9\frac{n(n+1)}{n^2}\right) = 24 - 9(1) = 13.$$

The Definite Integral

Concepts to Master

A. Definitions of $\int_a^b f(x)\,dx$, Reimann Sum; Integrability

B. Properties of definite integrals

Summary and Focus Questions

A. Given a partition P, a <u>Riemann sum</u> for a function f on $[a, b]$ is the number of the form

$$\sum_{i=1}^{n} f(x_i^*)\Delta x_i,$$

where each x_i^* is chosen from $[x_{i-1}, x_i]$, $i = 1, 2, 3, \ldots, n$.

The <u>definite integral of $f(x)$ from a to b</u>, denoted $\int_a^b f(x)\,dx$, is the limit of the values of the Riemann sums for partitions P whose norms approach 0.

This means $\int_a^b f(x)\,dx$ is a number such that for all $\varepsilon > 0$, there exists $\delta > 0$ such that if P is a partition of $[a, b]$ with $\|P\| < \delta$, then

$$\left|\sum_{i=1}^{n} f(x_i^*)\Delta x_i - \int_a^b f(x)\,dx\right| < \varepsilon$$ for any choice of numbers x_i^* in the

subintervals $[x_{i-1}, x_i]$ of P. When this limit exists we say that f is <u>integrable</u> on $[a, b]$. The definite integral always exists for continuous functions and monotonic functions on closed intervals.

For $f(x) \geq 0$ on $[a, b]$, $\int_a^b f(x)\,dx$ is the area under the graph of f between $x = a$ and $x = b$. For functions that are not nonnegative, the integral represents a kind of "net area".

For example, for the function graphed here, $\int_a^b f(x)\,dx = -A_1 + A_2 - A_3 + A_4$.

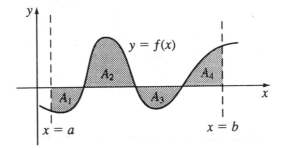

The simplest Riemann sum to calculate is the case where the partition P is regular (all Δx_i are equal to $\frac{b-a}{n}$) and $x_i^* = x_i$, the right endpoint of the ith subinterval. In this case the Riemann sum is:

$$\frac{b-a}{n} \sum_{i=1}^{n} f\left(a + i\frac{b-a}{n}\right).$$

Using midpoints for x_i^* and regular partitions, the Riemann sum is

$$\frac{b-a}{n} \sum_{i=1}^{n} f(\overline{x_i}), \text{ where } \overline{x_i} = \frac{x_i + x_{i-1}}{2}.$$

1) Construct a Riemann sum for $f(x) = x^2$ on the interval $[3,\,10]$ using the partition $[3,\,4]$, $[4,\,6]$, $[6,\,9]$, $[9,\,10]$.

We must select x_i^* points in $[x_{i-1},\,x_i]$ and calculate Δx_i for each subinterval:

Select $x_1^* = 3 \in [3,\,4];\ \Delta x_1 = 1$
$x_2^* = 5 \in [4,\,6];\ \Delta x_2 = 2$
$x_3^* = 8 \in [6,\,9];\ \Delta x_3 = 3$
$x_4^* = 9 \in [9,\,10];\ \Delta x_4 = 1$

(Your x_i^* choices may be different.)
The resulting Riemann sum for these choices of x_i^* is:

$$\sum_{i=1}^{4} f(x_i^*)\Delta x_i$$

$$= f(3) \cdot 1 + f(5) \cdot 2 + f(8) \cdot 3$$
$$+ f(9) \cdot 1$$
$$= (9)1 + (25)2 + (64)3 + (81)1$$
$$= 332$$

2) Does $\int_1^6 [\![x]\!]\, dx$ exist?

Yes, because $f(x) = [\![x]\!]$ is increasing (monotone) on $[1, 6]$.

3) Approximate $\int_1^4 x^3\, dx$ using a regular partition with $n = 3$ and midpoints for x_i^*.

$\frac{b-a}{n} = \frac{4-1}{3} = 1$.

The Rimann sum is calculated as follows:

i	subint	$\overline{x}_i$ mid pt	$f(\overline{x}_i)$
1	$[1, 2]$	1.5	3.375
2	$[2, 3]$	2.5	15.625
3	$[3, 4]$	3.5	42.875

$$\frac{b-a}{n}\sum_{i=1}^{3} f(\overline{x}_i) = 1(3.375 + 15.625$$
$$+ 42.875) = 61.875.$$

4) Write a definite integral for the shaded area.

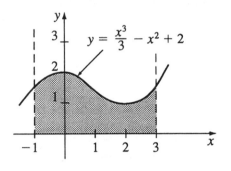

$$y = \frac{x^3}{3} - x^2 + 2$$

$\int_{-1}^{3} \left(\frac{x^3}{3} - x^2 + 2\right) dx$.

5) Suppose $f(x)$ is continuous and increasing on $[a, b]$. Let $\Re$ be the value of a Riemann sum for $\int_a^b f(x)\, dx$ using the left endpoint of each subinterval. Is $\Re$ greater than, equal to, or less than $\int_a^b f(x)\, dx$?

Less than, since the left endpoint will have the smallest value of $f(x)$ in each subinterval.

6) Evaluate $\int_3^6 (12 - x)dx$.
    (Hint: what area is it?)

The integral represents the trapezoid area below.

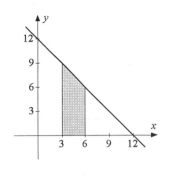

This area is $\frac{1}{2}(9 + 6)3 = 22.5$

B. Several integral properites allow definite integrals to be evaluated and estimated. If f and g are integrable functions on an interval containing a, b, and c,

$\int_a^b f(x)dx = -\int_b^a f(x)dx.$

$\int_a^b f(x)dx = \int_a^c f(x)dx + \int_c^b f(x)dx.$

$\int_a^b (f(x) \pm g(x))dx = \int_a^b f(x)dx \pm \int_a^b g(x)dx.$

$\int_a^b kf(x)dx = k\int_a^b f(x)dx$ (where k is a constant).

If $f(x) \geq g(x)$ for all $x \in [a, b]$, then $\int_a^b f(x)dx \geq \int_a^b g(x)dx.$

If $m \leq f(x) \leq M$ for $x \in [a, b]$, then $m(b - a) \leq \int_a^b f(x)dx \leq M(b - a).$

7) What are the missing limits of integration?
    $\int_4^? x^3 \, dx = \int_?^? x^3 \, dx + \int_5^7 x^3 \, dx$

$\int_4^7, \int_4^5, \int_5^7.$

8) True or False:

    a) $\int_0^1 (\sin x + \cos x)dx$
        $= \int_0^1 \sin x \, dx + \int_0^1 \cos x \, dx$

True.

b) $\int_0^1 x(x^2 + 1)dx$
$$= \left(\int_0^1 x\, dx\right)\left(\int_0^1 (x^2 + 1)dx\right)$$

False.

c) $\int_0^1 x(x^2 + 1)dx = x\int_0^1 (x^2 + 1)dx$

False.

d) $\int_0^1 x(t^2 + 1)dt = x\int_0^1 (t^2 + 1)dt$

True. (x is a constant here because t is the variable of integration.)

e) $\int_0^{\pi/4} \tan x\, dx \geq 0$

True, because for $0 \leq x \leq \frac{\pi}{4}$, $\tan x \geq 0$.

f) $\int_0^1 2^x\, dx \leq \int_0^1 x\, dx$

False, for $0 \leq x \leq 1$, $x < 2^x$, hence
$\int_0^1 x\, dx \leq \int_0^1 2^x\, dx$.

g) $\int_1^3 \frac{1}{x}\, dx = \int_1^5 \frac{1}{x}\, dx + \int_5^3 \frac{1}{x}\, dx$

True.

9) Using $1 \leq x^2 + 1 \leq 10$ for $1 \leq x \leq 3$ estimate $\int_1^3 (x^2 + 1)dx$.

Since $1 \leq x^2 + 1 \leq 10$ on $[1, 3]$,
$1(3 - 1) \leq \int_1^3 (x^2 + 1)dx \leq 10(3 - 1)$.
Thus $2 \leq \int_1^3 (x^2 + 1)dx \leq 20$.

The Fundamental Theorem of Calculus

Concepts to Master

A. Both forms of the Fundamental Theorem of Calculus

B. Evaluation of definite integrals using the Fundamental Theorem; Antiderivative properties

---

Summary and Focus Questions

A. The Fundamental Theorem of Calculus may be expressed in two different forms:

Part 1. If f is continuous on $[a, b]$ and if g is a function defined by
$g(x) = \int_a^x f(t)dt$, then g is continuous on $[a, b]$ and $g'(x) = f(x)$ for all $x \in (a, b)$.
For example, if $g(x) = \int_2^x (t^2 + 2t)dt$, then $g'(x) = x^2 + 2x$.

Part 2. If f is continuous on $[a, b]$ and F is an antiderivative of f, then
$\int_a^b f(x)dx = F(b) - F(a)$.
For example, since an antiderivative of $f(x) = 6x^2 + 2x$ is
$F(x) = 2x^3 + x^2$, we have
$\int_1^3 (6x^2 + 2x)dx = F(3) - F(1) = [2(3)^3 + 3^2] - [2(1)^3 + 1^2] = 60$.

1) If $h(x) = \int_4^x \sqrt{t}\, dt$, then
    $h'(x) = $ _____.

    $h'(x) = \sqrt{x}$. (Remember that h is a function of x, not t.

2) If $f(x)$ is continuous on $[a, b]$,
    then for all $x \in (a, b)$,
    $\frac{d}{dx}\int_a^x f(t)dt = $ _____.

    $f(x)$. This is Part 1 of the Fundamental Theorem of Calculus.

3) If $f'(x)$ is continuous on $[a, b]$,
    then for all $x \in [a, b]$,
    $\int_a^x f'(x)dx = $ _____.

$f(x) - f(a)$. This is the second part of the Fundamental Theorem of Calculus.

4) What does questions 2) and 3) say
    about differentiation and integration.

Question 2) says differentiation will
"undo" integration while 3) says
integration will "undo" taking a derivative.
Thus integration and differentiation are
inverse processes.

5) Evaluate $\int_0^1 (1 - x)dx$

An antiderivative of $1 - x$ is
$F(x) = x - \frac{x^2}{2}$.
Thus $\int_0^1 (1 - x)dx = F(1) - F(0)$
$= \left(1 - \frac{1}{2}\right) - \left(0 - \frac{0}{2}\right) = \frac{1}{2}$.

B. The notation for the antiderivative of f is $\int f(x)dx$, the indefinite integral of f
with respect to x.

For example, $\int 2x \, dx = x^2 + C$, where C is an arbitrary constant, the constant
of integration.

Here are some properties of antiderivatives:
  The derivative of $\int f(x)dx$ is $f(x)$, by definition.
  $\int c \, f(x)dx = c \int f(x)dx$ (c, a constant)
  $\int (f(x) \pm g(x))dx = \int f(x)dx \pm \int g(x)dx$
  $\int x^n dx = \frac{x^{n+1}}{n+1} + C$ for any number n except -1.
  $\int \sin x \, dx = -\cos x + C$ $\qquad$ $\int \sec x \tan x \, dx = \sec x + C$
  $\int \cos x \, dx = \sin x + C$ $\qquad$ $\int \csc^2 x \, dx = -\cot x + C$
  $\int \sec^2 x \, dx = \tan x + C$ $\qquad$ $\int \csc x \cot x \, dx = -\csc x + C$

Part 2 of the Fundamental Theorem of Calculus provides a handy procedure
for evaluating some integrals without resorting to limits of Riemann sums:
  1. Find $\int f(x)dx = F(x)$ (any antiderivative of f):
  2. Compute $F(x) \Big|_a^b = F(b) - F(a)$.

Remember that the method may be applied when f is continuous on $[a, b]$.

6) Evaluate

  a) $\int x^4 \, dx$

  b) $\int (6x + 12x^3) dx$

  c) $\int (2 \cos x - \sec^2 x) dx$

$\frac{x^5}{5} + C$

$3x^2 + 3x^4 + C$

$2 \sin x - \tan x + C$

7) Determine

  a) $\int_0^1 \sqrt{x} \, dx$

$\int_0^1 x^{1/2} \, dx = \frac{x^{3/2}}{\frac{3}{2}} = \frac{2}{3} x^{3/2} \Big|_0^1$

$\quad = \frac{2}{3}(1^{3/2} - 0^{3/2}) = \frac{2}{3}.$

  b) $\int_{-1}^3 x^{-3} \, dx$

The Fundamental Theorem does not apply because x^{-3} is not continuous at $x = 0$. We will have to wait until Chapter 7, Section 9, to evaluate this integral.

  c) $\int_{\pi/4}^{\pi/3} \sec x \tan x \, dx$

$\int_{\pi/4}^{\pi/3} \sec x \tan x \, dx = \sec x \Big|_{\pi/4}^{\pi/3}$

$\quad = \sec \frac{\pi}{3} - \sec \frac{\pi}{4} = 2 - \sqrt{2}.$

8) True or False:

  a) $\int 5f(x) \, dx = 5 \int f(x) dx.$

  b) $\int [f(x)]^2 \, dx = (\int f(x) dx)^2.$

  c) $\int f(x) g(x) dx$
$\quad = \int f(x) dx \int g(x) dx.$

True.

False.

False. (Try $f(x) = x^3$ and $g(x) = x^4$.)

125

The Substitution Rule

Concepts to Master

Integral evaluation using substitution

Summary and Focus Questions

The method of substitution is a process of rewriting an integral initially given in the form $\int f(g(x))g'(x)dx$ in the form $\int f(u)du$, where $u = g(x)$. It requires analyzing the given integral to determine a proper choice for u and the corresponding differential $du = g'(x)dx$.

For example, $\int (x^2 + 1)^3 x \, dx$ has the best choice $u = x^2 + 1$. For then the differential $du = 2x \, dx$ and thus $x \, dx = \frac{du}{2}$. Now make the substitutions:

$$\int (x^2 + 1)^3 x \, dx = \int u^3 \frac{du}{2} = \frac{1}{2} \int u^3 \, du = \frac{1}{2} \frac{u^4}{4} + C = \frac{(x^2+1)^4}{8} + C.$$

Frequently the integral will contain a multiple of $g'(x)dx$ with the final integral containing a constant (like the $\frac{1}{2}$ above).

In the case of a definite integral, such as $\int_0^1 (x^2 + 1)^3 x \, dx$, you may either
1. evaluate using the limits 0 and 2 after substitution:
$$\int_0^2 (x^2 + 1)^3 x \, dx = \frac{(x^2+1)^4}{8} \Big|_0^2 = \frac{625}{8} - \frac{1}{8} = 78$$

or

2. change the limits to fit u:
    Since $u(x) = x^2 + 1$, $u(2) = 5$ and $u(0) = 1$.
    Hence $\int_0^1 (x^2 + 1)^3 x \, dx = \frac{1}{2} \int_1^5 u^3 \, du = \frac{u^4}{8} \Big|_1^5 = \frac{625}{8} - \frac{1}{8} = 78$.

1) Use the substitution $u = x^2 + 3$ to evaluate $\int x\sqrt{x^2 + 3} \, dx$.

$u = x^2 + 3$, so $du = 2x \, dx$ and $x \, dx = \frac{du}{2}$.
$\int x(x^2 + 3)^{1/2} \, dx = \int u^{1/2} \frac{du}{2} = \frac{1}{2} \int u^{1/2} du$
$= \frac{1}{2} \cdot \frac{2}{3} u^{3/2} = \frac{1}{3} (x^2 + 3)^{3/2} + C$

2) State a substitution $u = g(x)$ that may be used for each of the following. Write the simplified integral and evaluate.

a) $\int (x^3 + 3x)^4 (x^2 + 1) dx$

$u =$ _____

$du =$ _____

$u = x^3 + 3x$

$du = (3x^2 + 3)dx = 3(x^2 + 1)dx.$

Thus $(x^2 + 1)dx = \frac{du}{3}$ and

$\int (x^3 + 3x)^4 (x^2 + 1) dx = \int u^4 \frac{du}{3}$

$\quad = \frac{1}{3}\int u^4 du = \frac{1}{3}\frac{u^5}{5} = \frac{1}{15}u^5$

$\quad = \frac{1}{15}(x^3 + 3x)^5 + C.$

b) $\int x^2 \sin x^3 \, dx$

$u =$ _____

$du =$ _____

$u = x^3$

$du = 3x^2 \, dx.$

Thus $x^2 \, dx = \frac{du}{3}$ and $\int x^2 \sin x^3 \, dx$

$\quad = \int \sin u \frac{du}{3} = \frac{1}{3}\int \sin u \, du$

$\quad = -\frac{1}{3}\cos u = -\frac{1}{3}\cos x^3 + C.$

c) $\int \tan^2 x \sec^2 x \, dx$

$u =$ _____

$du =$ _____

$u = \tan x$

$du = \sec^2 x \, dx.$

Thus $\int (\tan x)^2 \sec^2 x \, dx = \int u^2 \, du = \frac{1}{3}u^3$

$\quad = \frac{1}{3}\tan^3 x + C.$

3) Evaluate $\int_0^1 \sqrt{2x + 1} \, dx$

$u = 2x + 1, \, du = 2 \, dx, \, dx = \frac{du}{2}.$

Thus $\int_0^1 (2x + 1)^{1/2} \, dx = \int u^{1/2} \frac{du}{2}$

$\quad = \frac{1}{2}\int u^{1/2} \, du = \frac{1}{2} \cdot \frac{2}{3}u^{3/2}$

$\quad = \frac{1}{3}(2x + 1) \Big|_0^1 = \frac{1}{3}(3 - 1) = \frac{2}{3}.$

4) Evaluate by changing the limits of integration during substitution:
$\int_0^{\pi/6} (\cos^3 x + 1) \sin x \; dx$

$u = \cos x$, $du = -\sin x \; dx$, so $\sin x \; dx = -du$.
At $x = 0$, $u = \cos 0 = 1$.
At $x = \frac{\pi}{6}$, $u = \cos \frac{\pi}{6} = \frac{1}{2}$.
Therefore
$\int_0^{\pi/6} (\cos^3 x + 1) \sin x \; dx$
$= \int_1^{1/2} (u^3 + 1) \, -du$
$= -\int_1^{1/2} (u^3 + 1) du$
$= \int_{1/2}^1 (u^3 + 1) du = \left(\frac{u^4}{4} + u \right) \Big|_{1/2}^1$
$= \left(\frac{1}{4} + 1 \right) - \left(\frac{1}{64} + \frac{1}{2} \right) = \frac{47}{64}$.

5) Evaluate $\int_{-\pi/2}^{\pi/2} \sin x \; dx$

Since $f(x) = \sin x$ is symmetric about the origin, this integral is 0.

6) Evaluate $\int x^5 (x^3 + 2)^2 \; dx$

The first substitution to try is $u = x^3 + 2$, $du = 3x^2 \; dx$, and $x^2 \; dx = \frac{du}{3}$. Since we have an x^5 term it appears we may need a different substitution. But
$x^5 = x^3 \cdot x^2 = (u - 2)x^2$ since
$u = x^3 + 2$. So
$\int x^5 (x^3 + 2)^2 \; dx = \int (u - 2)u^2 \frac{du}{3}$
$= \frac{1}{3} \int (u^3 - 2u^2) du = \frac{1}{3} \left(\frac{u^4}{4} - \frac{2}{3} u^3 \right)$
$= \frac{u^4}{12} - \frac{2u^3}{9} = \frac{(x^3+2)^4}{12} - \frac{2(x^3+2)^3}{9} + C$

5

Applications of Integration

Cartoons courtesy of Sidney Harris. Used by permission.

Area between Curves

Concepts to Master

A. Area between $y = f(x)$, $y = g(x)$, $x = a$, $x = b$
B. Area between $x = f(y)$, $x = g(y)$, $y = c$, $y = d$
C. Area enclosed by two curves

Summary and Focus Questions

A. The area A of the region bounded by continuous $y = f(x)$ and $y = g(x)$ and $x = a$, $x = b$ with $f(x) \geq g(x)$ on $[a, b]$ may be approximated by a Riemann sum with terms $[f(x_i^*) - g(x_i^*)]\Delta x_i$. Thus the area is

$$A = \lim_{\|P\| \to 0} \sum_{i=1}^{n} [f(x_i^*) - g(x_i^*)]\Delta x_i$$
$$= \int_a^b [f(x) - g(x)]dx.$$

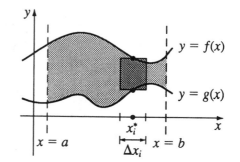

1) Set up a definite integral for the area of the region bounded by $y = x^3$, $y = 3 - x$, $x = 0$, $x = 1$.

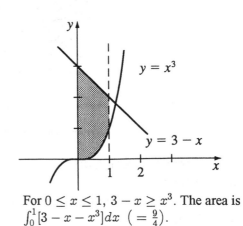

For $0 \leq x \leq 1$, $3 - x \geq x^3$. The area is $\int_0^1 [3 - x - x^3]dx$ $\left(= \frac{9}{4} \right)$.

B. The area of a region bounded by continuous $x = f(y)$, $x = g(y)$, $y = c$, $y = d$ with $f(y) \geq g(y)$ on $[c, d]$ is found in a manner similar to areas in part A. Here, however, x is a function of y and the area is

$$A = \lim_{\|P\| \to 0} \sum_{i=1}^{n} [f(y_i^*) - g(y_i^*)] \Delta y_i$$
$$= \int_c^d [f(y) - g(y)] dy.$$

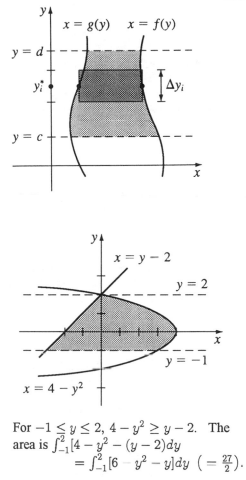

2) Set up a definite integral for the area of the region bounded by $x = 4 - y^2$, $x = y - 2$, $y = 2$, $y = -1$.

For $-1 \leq y \leq 2$, $4 - y^2 \geq y - 2$. The area is $\int_{-1}^{2} [4 - y^2 - (y - 2)] dy$
$$= \int_{-1}^{2} [6 - y^2 - y] dy \ \left(= \frac{27}{2} \right).$$

C. Some regions may be described so that either the $\int \ldots dx$ form or the $\int \ldots dy$ form may be used. You should select the form with the easier integral.

To find the area of regions such as the types sketched below, the x coordinates of the points of intersection a, b, c, $\ldots$ must be found. This is done by setting $f(x) = g(x)$ and solving for x. Often a sketch of the functions will help or a graphing calculator can be used to find approximate values for the points of intersection.

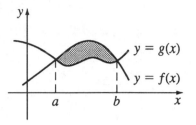

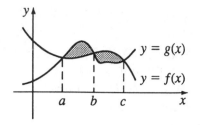

Area $= \int_a^b [f(x) - g(x)]dx$

Area $= \int_a^b [g(x) - f(x)]dx$
$\qquad + \int_b^c [f(x) - g(x)]dx$

3) Write an expression involving definite integrals for the area of each shaded region.

a)

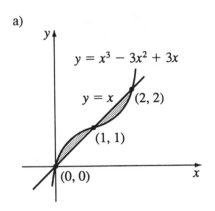

For $0 \le x \le 1$, $x^3 - 3x^2 + 3x \ge x$ while for $1 \le x \le 2$, $x \ge x^3 - 3x^2 + 3x$. Thus the area is the sum of two integrals:
$\int_0^1 [(x^3 - 3x^2 + 3x) - x]dx$
$\qquad + \int_1^2 (x - (x^3 - 3x^2 + 3x))dx$
$= \int_0^1 (x^3 - 3x^2 + 2x)dx$
$\qquad + \int_1^2 (-x^3 + 3x^2 - 2x)dx$
$\left(= \frac{1}{4} + \frac{1}{4} = \frac{1}{2} \right).$

b)

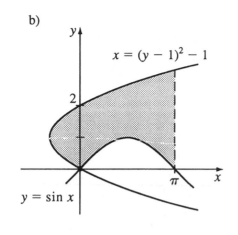

$x = (y-1)^2 - 1$

$y = \sin x$

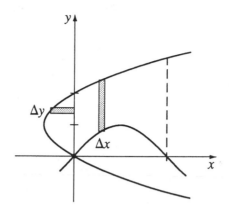

Δy

Δx

Using the y-axis to divide the region into two parts, the total area is two integrals. The area to the left of the y-axis is found using x as a function of y. The area to the right of the y-axis is found using y as a function of x: solving $x = (y-1)^2 - 1$ for y gives $y = 1 + \sqrt{x+1}$.

The total area is

$\int_0^2 -((y-1)^2 - 1)\,dy$
$\quad + \int_0^\pi \left((1 + \sqrt{x+1}) - \sin x\right)dx$
$= \int_0^2 (1 - (y-1)^2)\,dy$
$\quad + \int_0^\pi (1 + \sqrt{x+1} - \sin x)\,dx.$

4) Write an expression involving definite integrals for the area of the regions bounded by:

a) $y = 2x$ and $y = 8 - x^2$

$y = 2x$

-4

2

$y = 8 - x^2$

To find the points of intersection set
$2x = 8 - x^2$.
$x^2 + 2x - 8 = 0$
$(x-2)(x+4) = 0$, at $x = 2, -4$
The area is $\int_{-4}^2 (8 - x^2 - 2x)\,dx$ ($= 36$).

b) The shaded area between $y = \sin x$
    and $y = \cos x$.

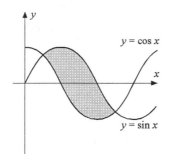

c) $y = x$, $y = -x$, $y = 2x - 3$

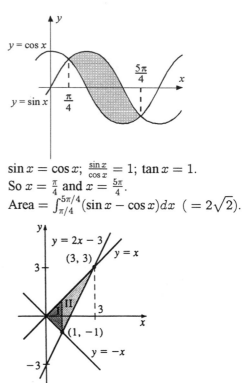

$\sin x = \cos x$; $\frac{\sin x}{\cos x} = 1$; $\tan x = 1$.
So $x = \frac{\pi}{4}$ and $x = \frac{5\pi}{4}$.
Area $= \int_{\pi/4}^{5\pi/4} (\sin x - \cos x)\, dx$ $(= 2\sqrt{2})$.

Divide the area into two regions:
Region I is bounded by $x = 0$, $x = 1$,
$y = x$, and $y = -x$.
Region II is bounded by $x = 1$, $x = 3$,
$y = x$, and $y = 2x - 3$.
The area is
$\int_0^1 [(x - (-x)]\, dx + \int_1^3 [x - (2x - 3)]\, dx$
$\qquad = \int_0^1 2x\ dx + \int_1^3 (3 - x)\, dx$ $(= 3)$.

Volumes

Concepts to Master

A. Determining volumes of solids by slicing
B. Determining volumes of solids of revolution (disks)
C. Determining volumes of solids of revolution (washers)

Summary and Focus Questions

A. Suppose the cross-sectional area of a solid cut by planes perpendicular to an x-axis is known to be $A(x)$ for each $x \in [a, b]$. The volume of a typical slice is approximately $A(x_i^*)\Delta x_i$ and thus the total volume obtained by this <u>method of slicing</u> is

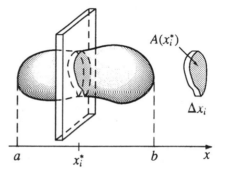

$$V = \lim_{\|P\| \to 0} \sum_{i=1}^{n} A(x_i^*)\Delta x_i = \int_a^b A(x)dx.$$

In many problems, the x-axis will have to be chosen carefully so that the function $A(x)$ can be determined.

1) Set up a definite integral for the volume of a right triangular solid that is 1 m, 1 m, and 2 m on its edges (see the figure).

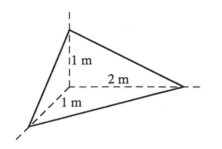

Draw an x-axis along the 2 meter side
with the origin in the corner.

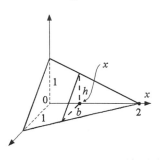

The area of the triangle determined by
slicing perpendicular to the x-axis at point
x is $A(x) = \frac{1}{2}bh$. By similar triangles,
$\frac{2-x}{2} = \frac{h}{1}$, so $h = 1 - \frac{x}{2}$. Likewise,
$b = 1 - \frac{x}{2}$, so $A(x) = \frac{1}{2}\left(1 - \frac{x}{2}\right)^2$.
The volume is $\int_0^2 \frac{1}{2}\left(1 - \frac{x}{2}\right)^2 dx \ \left(= \frac{1}{3}\right)$.

B. A <u>solid of revolution</u> is a solid
obtained by rotating a planar region
about a line. For a region bounded
by $y = f(x)$, $y = 0$, $x = a$, $x = b$,
and rotated about the x-axis, the
volume may be approximated by a
sum of volumes of disks. The
volume of a typical disk is
$\pi(\text{radius})^2(\text{thickness}) \approx \pi[f(x_i^*)]^2 \Delta x_i$.

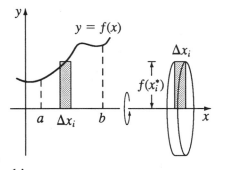

Thus the total volume by this <u>disk method</u> is
$$V = \lim_{\|P\| \to 0} \sum_{i=1}^{n} \pi[f(x_i^*)]^2 \Delta x_i = \int_a^b \pi[f(x)]^2 \, dx.$$

If the region is bounded on one side by a horizontal line $y = k$ and is rotated
about $y = k$, the volume is
$$V = \int_a^b \pi(k - f(x))^2 \, dx.$$

A solid obtained by rotating a region bounded by $x = g(y)$, $x = 0$, $y = c$,
$y = d$, about the y-axis has volume
$$V = \int_c^d \pi(g(y))^2 \, dy.$$

2) Set up a definite integral for the volume of each solid of revolution.

a) Rotate the region $y = 4 - x^2$, $y = 0$, $x = 0$, $x = 2$ about the x-axis.

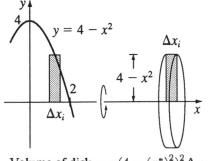

Volume of disk $= \pi(4 - (x_1^*)^2)^2 \Delta x_i$
Volume of solid $= \int_0^2 \pi(4 - x^2)^2 \, dx$
$= \left(\frac{256}{15}\pi\right)$.

b) The solid obtained by rotating about the y-axis the region bounded by $y = \sqrt{x}$, $x = 0$, $y = 4$.

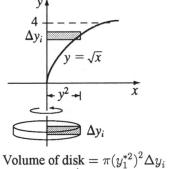

Volume of disk $= \pi(y_1^{*2})^2 \Delta y_i$
$= \pi(y_1^*)^4 \Delta y$
Volume of solid $= \int_0^4 \pi y^4 \, dy = \left(\frac{1024\pi}{5}\right)$.

c) The solid obtained by rotating about the line $y = 3$ the region bounded by $y = 10 - x^2$, $x = 0$, $x = 2$, $y = 3$.

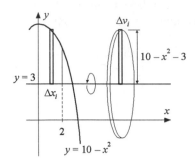

Volume of disk $= \pi(10 - (x_i^*)^2 - 3)^2 \Delta x_i$
Volume of solid $= \int_0^2 \pi(10 - x^2 - 3)^2 \, dx$

$$= \int_0^2 \pi(7 - x^2)^2 \, dx \quad \left(= \frac{2126}{15}\pi\right).$$

C. For a region bounded by $y = f(x)$,
$y = g(x)$ (with $f(x) \geq g(x)$),
$x = a$, $x = b$, and rotated about the
x-axis, the volume of the solid may
be approximated by a sum of
volumes of washers. The volume of
a typical washer is
$\pi(\text{outer radius})^2 (\text{thickness})$
$\quad - \pi(\text{inner radius})^2 (\text{thickness})$
$= \pi[(f(x_i^*))^2 - (g(x_i^*))^2]\Delta x_i$.
Thus the volume by this <u>washer method</u> is

$$V = \lim_{\|P\| \to 0} \sum_{i=1}^{n} \pi[(f(x_i^*))^2 - (g(x_i^*))^2]\Delta x_i = \int_a^b \pi[(f(x))^2 - (g(x))^2]dx.$$

Volumes of similar regions rotated about the y-axis are computed using
$$V = \int_c^d \pi[(f(y))^2 - (g(y))^2]dy.$$

3) Set up a definite integral for the
volume of the solid obtained by
rotation of the region between $x = 1$,
$x = 2$, $y = x^2$, and $y = x^3$ about the
x-axis.

Volumes by Cylindrical Shells

Concepts to Master

Volumes of solids of revolution by the shell method

Summary and Focus Questions

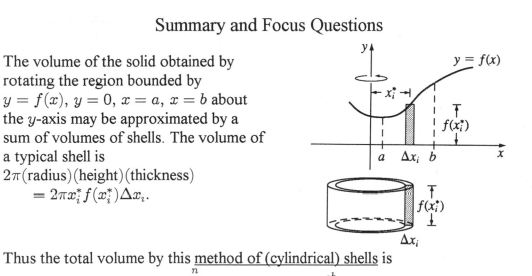

The volume of the solid obtained by rotating the region bounded by $y = f(x)$, $y = 0$, $x = a$, $x = b$ about the y-axis may be approximated by a sum of volumes of shells. The volume of a typical shell is
$2\pi(\text{radius})(\text{height})(\text{thickness})$
$\quad = 2\pi x_i^* f(x_i^*)\Delta x_i.$

Thus the total volume by this <u>method of (cylindrical) shells</u> is
$$V = \lim_{\|P\|\to 0} \sum_{i=1}^{n} 2\pi x_i^* f(x_i^*)\Delta x_i = \int_a^b 2\pi x f(x)\ dx.$$

Volumes of solids of revolution about the x-axis are computed similarly.

1) Set up a definite integral for the volume of the solid obtained by rotating about the y-axis the region bounded by $y = \sqrt{x}$, $y = 0$, $x = 1$, $x = 4$.

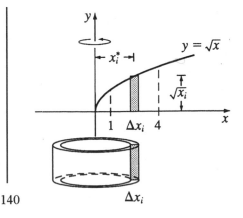

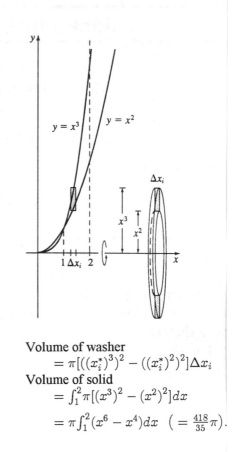

Volume of washer
$$= \pi[((x_i^*)^3)^2 - ((x_i^*)^2)^2]\Delta x_i$$
Volume of solid
$$= \int_1^2 \pi[(x^3)^2 - (x^2)^2]dx$$
$$= \pi\int_1^2 (x^6 - x^4)dx \quad \left(= \frac{418}{35}\pi\right).$$

Volume of shell $= 2\pi x_i^* \left(\sqrt{x_i^*} \right) \Delta x_i$
$$= 2\pi (x_i^*)^{3/2} \Delta x_i$$

Volume of solid $= \int_1^4 2\pi x^{3/2} \ dx$
$\left(= \frac{124\pi}{5} \right)$.

2) Set up a definite integral for the volume of the solid obtained by rotating about the x-axis the region bounded by $x = y^2$, $y = 0$, $x = 4$.

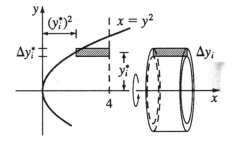

Volume of shell $= 2\pi y_i^* (4 - (y_i^*)^2) \Delta y_i$

$$= 2\pi (4y_i^* - (y_i^*)^3) \Delta y.$$

Volume of solid $= \int_0^2 2\pi (4y - y^3) dy$
$(= 56\pi)$.

Work

Concepts to Master

Work done by a varying amount of force

Summary and Focus Questions

Work is the measure of the total amount of effort, in units of foot-pounds or newton-meters (J, joules). With a constant force F the work done to move an object a distance d is $W = F \cdot d$.

If an object moves along an x-axis from a to b by a varying force $F(x)$, the work done is $W = \int_a^b F(x)dx$.

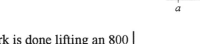

1) How much work is done lifting an 800 pound piano upward 6 feet?

   The force is a constant 800 pounds and the distance is 6 feet. The work is $W = (800)(6) = 4800$ ft-lb.

2) How much work is done moving a particle along an x-axis from 0 to $\frac{\pi}{2}$ feet if at each point x the amount of force is $\cos x$ pounds?

   $W = \int_0^{\pi/2} \cos x \; dx = \sin x \Big|_0^{\pi/2} = 1$ ft-lb.

3) Find an integral for how much work is done in pumping over the edge all the liquid of density ρ out of a 2 foot long trough in the shape of an equilateral triangle with sides of 1 foot.

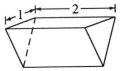

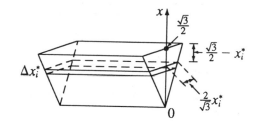

The work done to pump out a slice of the liquid is (density · gravitational constant · volume · height raised). The density is ρ; g is the gravitational constant. The volume of a slice is

$$\left(\tfrac{2}{\sqrt{3}}x_i^*\right)(2)(\Delta x_i) = \tfrac{4}{\sqrt{3}}x_i^* \Delta x_i.$$

The height raised is $\frac{\sqrt{3}}{2} - x_i$.

The work done lifting the slice is

$$\rho g\left(\tfrac{4}{\sqrt{3}}x_i^* \Delta x_i\right)\left(\tfrac{\sqrt{3}}{2} - x_i^*\right).$$

The total work is

$$\lim_{\|P\|\to 0} \sum_{i=1}^{n} \rho g\left(\tfrac{4}{\sqrt{3}}x_i^*\right)\left(\tfrac{\sqrt{3}}{2} - x_i^*\right)\Delta x_i$$

$$= \int_0^{\sqrt{3}/2} \rho g \tfrac{4}{\sqrt{3}}x\left(\tfrac{\sqrt{3}}{2} - x\right)dx$$

$$\left(= \tfrac{\rho g}{4}\right).$$

Average Value of a Function

Concepts to Master

A. Average value of a function
B. Mean Value Theorem for Integrals

Summary and Focus Questions

A. The <u>average value</u> of a continuous function $y = f(x)$ over a closed interval $[a, b]$ is $f_{ave} = \frac{1}{b-a} \int_a^b f(x) \, dx$.

This makes sense because we are taking the area $\int_a^b f(x) \, dx$ and dividing it by the "width" to obtain the (average) height of $f(x)$.

1) Find the average value of
   $f(x) = 2x + 6x^2$ over $[1, 4]$.

$$f_{ave} = \frac{1}{4-1} \int_1^4 (2x + 6x^2) dx$$
$$= \frac{1}{3}(x^2 + 2x^3) \Big|_1^4 = 47.$$

2) Find the average value of $\cos x$ over $[0, \pi]$.

$$\frac{1}{\pi - 0} \int_0^\pi \cos x \, dx = \frac{1}{\pi} \sin x \Big|_0^\pi$$
$$= \frac{1}{\pi}(0 - 0) = 0.$$

Intuitively this makes sense because the "right half" of the cosine curve (the part between $\frac{\pi}{2}$ and π) is below the x-axis and is symmetric with the other half above the x-axis. The cosine values average zero.

B. <u>The Mean Value Theorem for Integrals</u>:
   If f is continuous on a closed interval $[a, b]$ there exists $c \in [a, b]$ such that
   $\int_a^b f(x) \, dx = f(c)(b - a)$.

3) Mark on the graph a number c guaranteed by the Mean Value Theorem for $f(x)$:

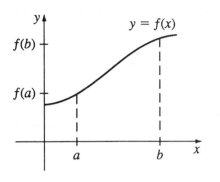

4) Let $F'(x) = f(x)$ be continuous on $[a, b]$. State the conclusion of the Mean Value Theorem for Integrals for f in terms of the function $F(x)$. Use the Fundamental Theorem of Calculus.

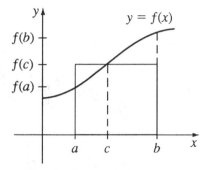

The number c requires $\int_a^b f(x)dx = f(c)(b - a)$ so the area under $f(x)$ must equal the area of the rectangle with width $b - a$ and height $f(c)$.

The conclusion is "there is a number c such that $\int_a^b f(x)dx = f(c)(b - a)$." By the Fundamental Theorem $\int_a^b f(x)dx = F(b) - F(a)$. Also $f(c) = F'(c)$. So the conclusion, in terms of $F(x)$, becomes "there is a number $c \in (a, b)$ such that $F(b) - F(a) = F'(c)(b - a)$" - precisely the Mean Value Theorem of Section 3.2!

5) For $f(x) = x^2$ on $[1, 4]$, find a value for c in the Mean Value Theorem for Integrals.

$\int_1^4 x^2 \, dx = \frac{x^3}{3} \Big|_1^4 = 21.$
So $21 = f(c)(4 - 1)$.
$f(c) = 7, c^2 = 7, c = \sqrt{7}.$

"IF IT'S TRUE THAT THE WORLD ANT POPULATION IS 10^{15} THEN IT'S NO WONDER WE NEVER RUN INTO ANYONE WE KNOW."

Cartoons courtesy of Sidney Harris. Used by permission.

Inverse Functions

Concepts to Master

A. One-to-one functions; Inverse of a function
B. Derivative of the inverse function

Summary and Focus Questions

A. A function f is one-to-one means that for all x_1, x_2 in the domain, if $x_1 \neq x_2$, then $f(x_1) \neq f(x_2)$.

For example, $f(x) = 2x + 5$ is one-to-one because if $x_1 \neq x_2$, $2x_1 + 5 \neq 2x_2 + 5$, while $f(x) = x^2$ is not one-to-one ($4 \neq -4$), yet $f(4) = 4^2 = (-4)^2 = f(-4)$).

Increasing and decreasing functions are one-to-one. If f is one-to-one with domain A and range B then the inverse function f^{-1} has domain B and range A, and

$$f^{-1}(b) = a \text{ iff } f(a) = b.$$

Thus a point (a, b) is on the graph of f if and only if (b, a) is on the graph of f^{-1}. These inverse properties hold

$$f^{-1}(f(x)) = x \text{ for all } x \text{ in the domain of } f$$
$$f(f^{-1}(x)) = x \text{ for all } x \text{ in the domain of } f^{-1}$$

To find the rule for $y = f^{-1}(x)$,
(1) solve the equation $y = f(x)$ for x.
(2) interchange x and y.

1) Is $f(x) = \sin x$ one-to-one?

No. For example, $0 \neq \pi$, yet $\sin 0 = 0 = \sin \pi$.

2) Is the graph below the graph of a function with an inverse function?

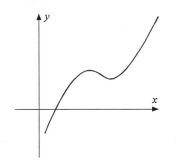

No. The function is not one-to-one.

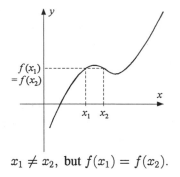

$x_1 \neq x_2$, but $f(x_1) = f(x_2)$.

3) Given the following graph of f, sketch the graph of f^{-1} on the same axis.

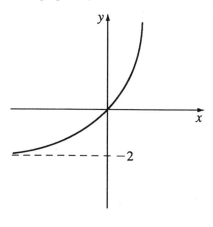

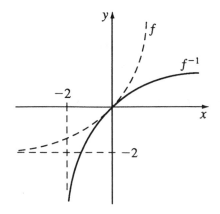

4) If f^{-1} is a function and $f(5) = 8$,
then $f^{-1}(8) =$ _____.

5.

5) Write the inverse function of
$f(x) = 6x + 30$.

(1) $y = 6x + 30$
$y - 30 = 6x$
$x = \frac{1}{6}y - 5$
(2) $y = \frac{1}{6}x - 5$, so $f^{-1}(x) = \frac{1}{6}x - 5$.

6) True, False:
If f is one-to-one with domain A and
range B, then $f(f^{-1}(x)) = x$ for all
$x \in A$.

False. $f(f^{-1}(x)) = x$ for all $x \in \underline{B}$.

B. If f is one-to-one, differentiable, and $f(b) = a$, then $(f^{-1})'(a) = \frac{1}{f'(b)}$,
provided $f'(b) \neq 0$.

Since $f^{-1}(a) = (b)$, this may also be written as:
$$(f^{-1})'(a) = \frac{1}{f'(f^{-1}(a))}.$$

7) Suppose $f(3) = 5$, $f'(3) = 6$ and f is
one-to-one. Then $(f^{-1}(5))' =$ _____.

$\frac{1}{f'(3)} = \frac{1}{6}$.

8) Let $f(x) = x^3 + 2x + 3$.
Find $(f^{-1})'(3)$.

We first must find b such that $b = f^{-1}(3)$;
that is, $f(b) = 3$.
$f(b) = b^3 + 2b + 3 = 3$
$b^3 + 2b = 0$
$b(b^2 + 2) = 0$.
Since $b^2 + 2 \neq 0$, $b = 0$.
$f'(x) = 3x^2 + 2$, so
$f'(0) = 3(0^2) + 2 = 2$.
Hence $(f^{-1})'(3) = \frac{1}{f'(0)} = \frac{1}{2}$.

9) Given the graph of f and that the
tangent line to f at $(4, 7)$ has slope 3,
draw the graph of f^{-1} and indicate the
slope of the tangent to f^{-1} at $(7, 4)$.

Properties of exponentials include:

$$a^{x+y} = a^x a^y \qquad a^{x-y} = \frac{a^x}{a^y} \qquad a^{xy} = (a^x)^y \qquad (ab)^x = a^x b^x$$

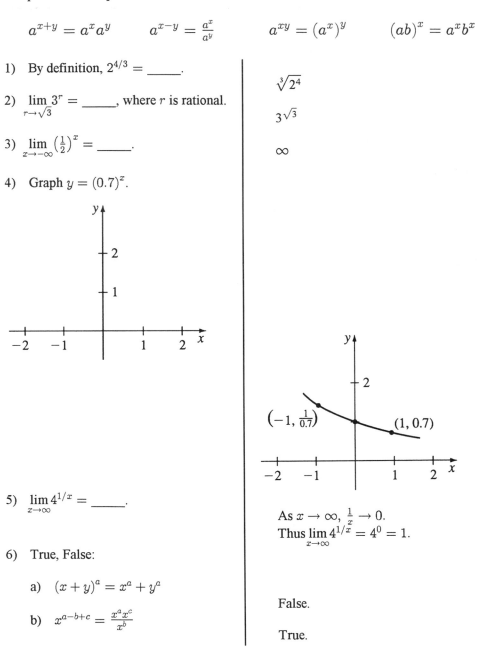

1) By definition, $2^{4/3} = \underline{\hspace{1cm}}$.

$\sqrt[3]{2^4}$

2) $\lim\limits_{r \to \sqrt{3}} 3^r = \underline{\hspace{1cm}}$, where r is rational.

$3^{\sqrt{3}}$

3) $\lim\limits_{x \to -\infty} \left(\frac{1}{2}\right)^x = \underline{\hspace{1cm}}$.

∞

4) Graph $y = (0.7)^x$.

$\left(-1, \frac{1}{0.7}\right)$ $(1, 0.7)$

5) $\lim\limits_{x \to \infty} 4^{1/x} = \underline{\hspace{1cm}}$.

As $x \to \infty$, $\frac{1}{x} \to 0$.
Thus $\lim\limits_{x \to \infty} 4^{1/x} = 4^0 = 1$.

6) True, False:

a) $(x + y)^a = x^a + y^a$

False.

b) $x^{a-b+c} = \frac{x^a x^c}{x^b}$

True.

B. The number e is the constant (approximately 2.718) such that $\lim\limits_{h \to 0} \frac{e^h - 1}{h} = 1$.

The function $f(x) = e^x$ has the usual exponential function properties: continuous, increasing, concave upward,

$\lim\limits_{x \to \infty} e^x = \infty$,

$\lim\limits_{x \to -\infty} e^x = 0$.

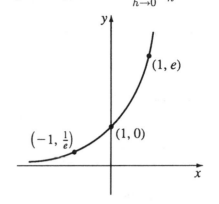

7) For what number(s) k is $\lim\limits_{h \to 0} \frac{k^h - 1}{h} = 1$?

$k = e$ is the only such number.

8) $\lim\limits_{x \to \infty} e^{-x}$

As $x \to \infty$, $-x \to -\infty$.
Thus $\lim\limits_{x \to \infty} e^{-x} = 0$.

C. In general, for $f(x) = a^x$, $f'(x) = f'(0) \cdot a^x$; that is, $(a^x)' = a^x \cdot \left(\lim\limits_{h \to 0} \frac{a^h - 1}{h} \right)$.

Thus the number e was chosen in order to make this differentiation formula simple:

$$\text{If } f(x) = e^x, \text{ then } f'(x) = e^x.$$

$\frac{de^{u(x)}}{dx} = e^{u(x)} u'(x)$.

For example, if $f(x) = e^{3x^2}$, then $f'(x) = e^{3x^2}(6x)$.

$\int e^x \, dx = e^x + C.$

9) Find $f'(x)$ for:

a) $f(x) = e^{\sqrt{x}}$

$f(x) = e^{x^{1/2}}$,
$f'(x) = e^{x^{1/2}} \left(\frac{1}{2} x^{-1/2} \right) = \frac{e^{\sqrt{x}}}{2\sqrt{x}}$.

b) $f(x) = e^3$

$f'(x) = 0$. (Remember, e is a constant and therefore e^3 is a constant.)

c) $f(x) = x^2 \cos e^x$

$$f'(x) = x^2(-\sin e^x e^x) + \cos e^x (2x)$$
$$= -x^2 e^x \sin e^x + 2x \cos e^x.$$

10) For what a is $(a^x)' = a^x$?

Only for $a = e$.

11) Evaluate

   a) $\int x e^{2x^2}\, dx$

Let $u = 2x^2$, $du = 4x\, dx$.
$x\, dx = \frac{du}{4}$
$\int x e^{2x^2}\, dx = \int e^u \frac{du}{4} = \frac{1}{4}\int e^u\, du$
$$= \frac{1}{4} e^u = \frac{1}{4} e^{2x^2} + C.$$

   b) $\int_0^1 e^x \sin e^x\, dx$

Let $u = e^x$, $du = e^x\, dx$
$\int_0^1 e^x \sin e^x\, dx = \int \sin u\, du$
$$= -\cos u = -\cos e^x \Big|_0^1$$
$$= -\cos e^1 - (-\cos e^0)$$
$$= \cos 1 - \cos e.$$

12) Simplify to e to a power:
   $\frac{(e^x \cdot e^3)^2}{e}$

$$\frac{(e^x \cdot e^3)^2}{e} = \frac{(e^{x+3})^2}{e} = \frac{e^{2x+6}}{e}$$
$$= e^{2x+6-1} = e^{2x+5}.$$

Logarithmic Functions

Concepts to Master

Definition, properties and graph of $y = \log_a x$.

Summary and Focus Questions

$\log_a x = y$ means $a^y = x$.
This definition requires $x > 0$, since $a^y > 0$ for all y. $\log_a x$ is the exponent you put on a to get x.

The logarithmic function $y = \log_a x$ is defined as the inverse of the exponential $y = a^x$. Thus

$$\log_a(a^x) = x$$
$$a^{\log_a x} = x, \text{ for } x > 0$$
$$\log_a(xy) = \log_a x + \log_a y$$
$$\log_a\left(\frac{x}{y}\right) = \log_a x - \log_a y$$
$$\log_a x^c = c \log_a x$$
$$\lim_{x \to \infty} \log_a x = \infty$$
$$\lim_{x \to 0^+} \log_a x = -\infty$$

For $a > 1$, the graph of $y = \log_a x$ is

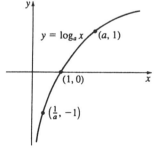

In the special case that $a = e$, $\underline{\ln x}$ means $\log_a x$.
Thus $\ln e = \log_e e = 1$.
$\log_a x = \frac{\ln x}{\ln a}$.

B. $\int \frac{1}{x} \, dx = \ln|x| + C$.

This completes the power rule for integrals:

$$\int x^n \, dx = \begin{cases} \frac{x^{n+1}}{n+1}, & \text{for } n \neq -1 \\ \ln|x|, & \text{for } n = -1 \end{cases}$$

$\int a^x \, dx = \frac{a^x}{\ln a} + C$

2) Evaluate each:

a) $\int_1^3 \frac{1}{x} \, dx$

$\int_1^3 \frac{1}{x} \, dx = \ln|x| \, \Big|_1^3 = \ln 3 - \ln 1$
$\qquad = \ln 3 - 0 = \ln 3$.

b) $\int \frac{x^3}{2+x^4} \, dx$

Let $u = 2 + x^4$, $du = 4x^3 \, dx$
so $x^3 \, dx = \frac{du}{4}$.
$\int \frac{x^3}{2+x^4} \, dx = \int \frac{1}{2+x^4} x^3 \, dx$

$\qquad = \int \frac{1}{u} \frac{du}{4} = \frac{1}{4} \int \frac{1}{u} \, du$
$\qquad = \frac{1}{4} \ln|u|$
$\qquad = \frac{1}{4} \ln(2+x^4)$
$\qquad = \ln \sqrt[4]{2+x^4} + C$.

Note that absolute value signs are not needed because $2 + x^4 > 0$ for all x.

3) $\int 7^{-x} \, dx$

$u = -x$, $du = -dx$, $dx = -du$
$\int 7^{-x} \, dx = \int 7^u(-du) = -\int 7^u \, du$
$\qquad = -\frac{7^u}{\ln 7} = -\frac{7^{-x}}{\ln 7} + C$.

C. The derivative of a function $y = f(x)$ may be found using logarithmic differentiation, that is,

1. Take ln of both sides of $y = f(x)$: $\ln y = \ln(f(x))$.
2. Simplify $\ln(f(x))$ using logarithm properties.
3. Differentiate both sides implicitly and solve for y' : $y' = y \frac{d}{dx} \ln(f(x))$.

4) Find y' for each, using logarithmic differentiation.

a) $y = \left(\frac{10x^3}{\sqrt{x+1}}\right)^4$

$\ln y = \ln\left(\frac{10x^3}{\sqrt{x+1}}\right)^4$

$\qquad = 4\left(\ln 10x^3 - \ln\sqrt{x+1}\right)$

$\qquad = 4\left(\ln 10 + 3\ln x - \frac{1}{2}\ln(x+1)\right)$

$\frac{d}{dx}\ln y = 4\left(0 + \frac{3}{x} - \frac{1}{2}\frac{1}{x+1}(1)\right)$

$\qquad = 4\left(\frac{3}{x} - \frac{1}{2(x+1)}\right).$

Thus $y' = \left(\frac{10x^3}{\sqrt{x+1}}\right)^4\left[4\left(\frac{3}{x} - \frac{1}{2(x+1)}\right)\right].$

b) $y = 6^x$

$\ln y = \ln 6^x = x\ln 6.$

Thus $\frac{d}{dx}\ln y = \ln 6$ (a constant)

and $y' = y\ln 6 = 6^x\ln 6.$

5) Sketch a graph of $f(x) = x - \ln x$.

a.   The domain is $(0, \infty)$.

b. $x - \ln x$ is never 0, so no x-intercept.

c.   Since $x > 0$ there is no symmetry.

d. $\lim\limits_{x\to 0^+}(x - \ln x) = \infty$

$\lim\limits_{x\to\infty}(x - \ln x) = \infty$

e. $f'(x) = 1 - \frac{1}{x}.$

For $0 < x < 1$, $f'(x) < 0$ and f is decreasing.

For $x > 1$, $f'(x) > 0$ and f is increasing.

f. $f'(x) = 1 - \frac{1}{x} = 0$ at $x = 1.$

$f''(x) = x^{-2}$, so $f''(x) = 1 > 0.$

Thus f has a local minimum at $x = 1.$

g.   Since $f''(x) > 0$ the graph is always concave upward.

h.   Here is a sketch:

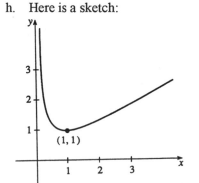

$(1, 1)$

D. $\lim_{x \to 0^+} (1 + x)^{1/x} = e.$

Equivalently, $\lim_{t \to \infty} \left(1 + \frac{1}{t}\right)^t = e.$

6) Find each limit:

a) $\lim_{h \to 0^+} (1 + h)^{1/h}$

e

b) $\lim_{x \to \infty} \left(1 + \frac{1}{x}\right)^{3x}$

$$\lim_{x \to \infty} \left(1 + \frac{1}{x}\right)^{3x} = \lim_{x \to \infty} \left[\left(1 + \frac{1}{x}\right)^x\right]^3$$
$$= \left[\lim_{x \to \infty} \left(1 + \frac{1}{x}\right)^x\right]^3 = e^3.$$

c) $\lim_{x \to 1^+} x^{1/(x-1)}$

Let $t = x - 1$, hence $x = 1 + t$.
As $x \to 1^+$, $t \to 0^+$. Thus
$$\lim_{x \to 1^+} x^{1/(x-1)} = \lim_{t \to 0^+} (1 + t)^{1/t} = e.$$

The Natural Logarithm Function

Concepts to Master

A. Integral definition of $\ln x$; properties of $\ln$; definition of e

B. Derivativeof $\ln x$; integral of $\frac{1}{x}$

C. Logarithmic differentiation

Summary and Focus Questions

This section and sections 6.3* and 6.4* provide an alternate means for defining logarithm and exponential functions and verifying their properties. You should choose to do these three sections or sections 6.2, 6.3, and 6.4.

A. The <u>natural logarithm function</u> is defined by
$$\ln x = \int_1^x \tfrac{1}{t} \, dt \text{ for } x > 0.$$

The function $y = \ln x$ has these properties:
domain of ln is $(0, \infty)$
range of ln is all reals
$\ln 1 = 0$
$\ln (xy) = \ln x + \ln y$
$\ln \left(\frac{x}{y} \right) = \ln x - \ln y$
$\ln (x^r) = r \ln x$
$\lim\limits_{x \to \infty} \ln x = \infty$
$\lim\limits_{x \to 0^+} \ln x = -\infty$
ln is increasing, concave down, and one-to-one.

e is the unique number for which $\ln e = 1$.

161

1) Find an expression for the shaded area:

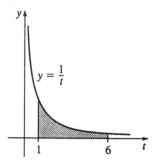

The shaded area is $\int_1^6 \frac{1}{t}\, dt$ which is $\ln 6$.

2) True or False:
   a) $\ln e = 1$

   True.

   b) $\ln(a - b) = \ln a - \ln b$

   False.

   c) $\ln\left(\frac{a}{b}\right) = -\ln\left(\frac{b}{a}\right)$

   True.

3) Write $2\ln x + 3\ln y$ as a single logarithm.

$$2\ln x + 3\ln y = \ln x^2 + \ln y^3$$
$$= \ln(x^2 y^3).$$

4) Expand to eliminate radicals and fractions:
   $\ln \sqrt[3]{\frac{x+7}{x^2}}$

$$\ln \sqrt[3]{\tfrac{x+7}{x^2}} = \ln \left(\tfrac{x+7}{x^2}\right)^{1/3} = \tfrac{1}{3}\ln\left(\tfrac{x+7}{x^2}\right)$$
$$= \tfrac{1}{3}[\ln(x+7) - \ln x^2]$$
$$= \tfrac{1}{3}[\ln(x+7) - 2\ln x]$$
$$= \tfrac{1}{3}\ln(x+7) - \tfrac{2}{3}\ln x.$$

5) Find the number k on the t-axis such that:
   a) the shaded area is 1.

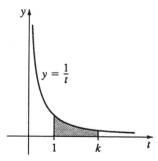

The shaded area is $\int_1^k \frac{1}{t}\, dt = \ln k = 1$, hence $k = e$.

b) the shaded area is 1.

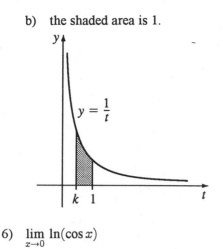

$y = \frac{1}{t}$

The area $= \int_k^1 \frac{1}{t}\, dt = \ln 1 - \ln k$
$= -\ln k = 1.$
Thus $\ln \frac{1}{k} = 1$, so $\frac{1}{k} = e$ or $k = \frac{1}{e}$.

6) $\lim_{x \to 0} \ln(\cos x)$

As $x \to 0$, $\cos x \to 1$.
Thus $\ln(\cos x) \to \ln(1) = 0.$

B. If $f(x) = \ln x$, $f'(x) = \frac{1}{x}$. In general $\frac{d \ln u(x)}{dx} = \frac{1}{u(x)} \cdot u'(x)$
$\int \frac{1}{x}\, dx = \ln|x| + C.$
This completes the power rule for integrals:
$$\int x^n \, dx = \begin{cases} \frac{x^{n+1}}{n+1}, & \text{for } n \neq -1 \\ \ln|x|, & \text{for } n = -1 \end{cases}$$

7) Find $f'(x)$ for each:

a) $f(x) = \ln x^2$

$f(x) = 2\ln x$, so $f'(x) = 2\left(\frac{1}{x}\right) = \frac{2}{x}$.
Alternatively, using the Chain Rule,
$f'(x) = \frac{1}{x^2}(2x) = \frac{2}{x}$.

b) $f(x) = \ln \cos x$

$f'(x) = \frac{1}{\cos x} \cdot (-\sin x) = -\tan x.$

c) $f(x) = \ln \sqrt{\frac{x^2+1}{2x^3}}$ $(x > 0)$

You should simplify $f(x)$ first:
$f(x) = \frac{1}{2} \ln \left(\frac{x^2+1}{2x^3} \right)$
$= \frac{1}{2}(\ln(x^2 + 1) - \ln 2 - 3 \ln x).$
$f'(x) = \frac{1}{2}\left(\frac{2x}{x^2+1} - \frac{3}{x} \right).$

d) $f(x) = x^2 \ln x$

$f'(x) = x^2 \left(\frac{1}{x} \right) + \ln x (2x)$
$= x + 2x \ln x$

8) Evaluate each:

   a) $\int_1^3 \frac{1}{x}\, dx$

   b) $\int \frac{x^3}{2+x^4}\, dx$

$\int_1^3 \frac{1}{x}\, dx = \ln|x|\Big|_1^3 = \ln 3 - \ln 1$
$$= \ln 3 - 0 = \ln 3.$$

Let $u = 2 + x^4$, $du = 4x^3\, dx$
so $x^3\, dx = \frac{du}{4}$.
$\int \frac{x^3}{2+x^4}\, dx = \int \frac{1}{2+x^4}x^3\, dx$

$$= \int \frac{1}{u}\frac{du}{4} = \frac{1}{4}\int \frac{1}{u}\, du = \frac{1}{4}\ln|u|$$

$$= \frac{1}{4}\ln(2+x^4) = \ln \sqrt[4]{2+x^4} + C.$$

Note that absolute value signs are not
needed because $2 + x^4 > 0$ for all x.

9) Sketch a graph of $f(x) = x - \ln x$.

a. The domain is $(0, \infty)$.
b. $x - \ln x$ is never 0, so no x-intercept.
c. Since $x > 0$ there is no symmetry.
d. $\lim\limits_{x \to 0^+} x - \ln x = \infty$
   $\lim\limits_{x \to \infty} x - \ln x = \infty.$
e. $f'(x) = 1 - \frac{1}{x}$.
   For $0 < x < 1$, $f'(x) < 0$ and f is
   decreasing.
   For $x > 1$, $f'(x) > 0$ and f is increasing.
f. $f'(x) = 1 - \frac{1}{x} = 0$ at $x = 1$.
   $f''(x) = x^{-2}$, so $f''(1) = 1 > 0$.
   Thus f has a local minimum at $x = 1$.
g. Since $f''(x) > 0$ the graph is always
   concave upward.
h. Here is a sketch:

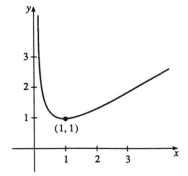

(1, 1)

C. The derivative of a function $y = f(x)$ may be found using logarithmic differentiation, that is,

1. Take ln of both sides of $y = f(x)$: $\ln y = \ln(f(x))$.
2. Simplify $\ln(f(x))$ using logarithm properties.
3. Differentiate both sides implicitly and solve for y': $y' = y\frac{d}{dx}\ln(f(x))$.

10) Find y', using logarithmic differentiation.

$$y = \left(\frac{10x^3}{\sqrt{x+1}}\right)^4$$

$$\ln y = \ln\left(\frac{10x^3}{\sqrt{x+1}}\right)^4$$
$$= 4\left(\ln 10x^3 - \ln\sqrt{x+1}\right)$$
$$= 4\left(\ln 10 + 3\ln x - \tfrac{1}{2}\ln(x+1)\right)$$
$$\tfrac{d}{dx}\ln y = 4\left(0 + \tfrac{3}{x} - \tfrac{1}{2}\tfrac{1}{x+1}(1)\right)$$
$$= 4\left(\tfrac{3}{x} - \tfrac{1}{2(x+1)}\right).$$

Thus $y' = \left(\frac{10x^3}{\sqrt{x+1}}\right)^4\left[4\left(\tfrac{3}{x} - \tfrac{1}{2(x+1)}\right)\right]$.

The Natural Exponential Function

Concepts to Master

A. Definition and properties of the natural exponential function
B. Derivative and integral of e^x

Summary and Focus Questions

A. The <u>natural exponential function</u>, $y = \exp(x)$ is defined as the inverse of $y = \ln x$. Thus $\exp(a) = b$ if and only if $\ln b = a$.

Recalling that e is a constant, approximately 2.718, it turns out that $\exp(x) = e^x$ for all x. Thus
$$e^a = b \text{ if and only if } \ln b = a.$$

The function $y = e^x$ has these properties:
domain of e^x is all reals
range of e^x is $(0, \infty)$
$e^{\ln x} = x$ for all $x > 0$.
$\ln e^x = x$ for all x.
$e^0 = 1$
$e^{x+y} = e^x e^y$
$e^{x-y} = \frac{e^x}{e^y}$
$(e^x)^r = e^{rx}$
$\lim\limits_{x \to \infty} e^x = \infty$
$\lim\limits_{x \to -\infty} e^x = 0$
$y = e^x$ is increasing, concave up, and one-to-one.

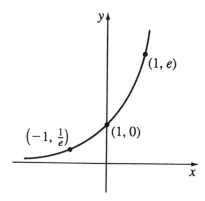

1) Simplify to e to a power:
$\frac{(e^x \cdot e^3)^2}{e}$

$\frac{(e^x \cdot e^3)^2}{e} = \frac{(e^{x+3})^2}{e} = \frac{e^{2x+6}}{e}$

$= e^{2x+6-1} = e^{2x+5}$.

2) $\lim\limits_{x\to\infty} e^{-x}$

As $x \to \infty$, $-x \to -\infty$.
Thus $\lim\limits_{x\to\infty} e^{-x} = 0$.

3) True, False:
    $e^{\ln x} = x$ for all x

False. This is only true for $x > 0$.

B. If $f(x) = e^x$, then $f'(x) = e^x$. In general $\dfrac{de^{u(x)}}{dx} = e^{u(x)}u'(x)$.
For example, if $f(x) = e^{3x^2}$, then $f'(x) = e^{3x^2}(6x)$.

$\int e^x \, dx = e^x + C$.

4) Find $f'(x)$ for:

a) $f(x) = e^{\sqrt{x}}$

$f(x) = e^{x^{1/2}}$,
$f'(x) = e^{x^{1/2}}\left(\frac{1}{2}x^{-1/2}\right) = \frac{e^{\sqrt{x}}}{2\sqrt{x}}$.

b) $f(x) = e^3$

$f'(x) = 0$. (Remember, e is a constant and therefore e^3 is a constant.)

c) $f(x) = x^2 \cos e^x$

$f'(x) = x^2(-\sin e^x e^x) + \cos e^x(2x)$
$= -x^2 e^x \sin e^x + 2x \cos e^x$.

5) Evaluate

a) $\int xe^{2x^2} \, dx$

Let $u = 2x^2$, $du = 4x \, dx$.
$x \, dx = \frac{du}{4}$
$\int xe^{2x^2} \, dx = \int e^u \frac{du}{4} = \frac{1}{4}\int e^u \, du$
$= \frac{1}{4}e^u = \frac{1}{4}e^{2x^2} + C$.

b) $\int_0^1 e^x \sin e^x \, dx$

Let $u = e^x$, $du = e^x \, dx$
$\int_0^1 e^x \sin e^x \, dx = \int_0^1 \sin u \, du$
$= -\cos u = -\cos e^x \Big|_0^1$
$= -\cos e^1 - (-\cos e^0)$
$= \cos 1 - \cos e$.

General Logarithmic and Exponential Function

Concepts to Master

A. Definition and properties of $y = a^x$; derivative and integral of a^x
B. Definition and properties of $y = \log_a x$; derivative of $\log_a x$
C. The number e as a limiting value

Summary and Focus Questions

A. a^x for $a > 0$ is defined as $a^x = e^{x \ln a}$. So, for example, $2^\pi = e^{\pi \ln 2}$.
Properties of exponentials include:

$$a^{x+y} = a^x a^y$$
$$a^{x-y} = \frac{a^x}{a^y}$$
$$a^{xy} = (a^x)^y$$
$$(ab)^x = a^x b^x$$

$f(x) = a^x$ has domain all reals, range $(0, \infty)$, is continuous and concave up.
The graph of $y = a^x$ is:

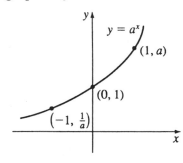

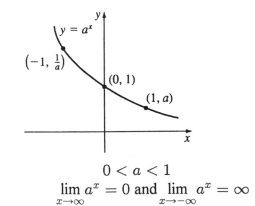

$a > 1$

$\lim\limits_{x \to \infty} a^x = \infty$ and $\lim\limits_{x \to -\infty} a^x = 0$

$0 < a < 1$

$\lim\limits_{x \to \infty} a^x = 0$ and $\lim\limits_{x \to -\infty} a^x = \infty$

If $f(x) = a^x$, $f'(x) = a^x \ln a$.
$\int a^x \, dx = \frac{a^x}{\ln a} + C$.

1) True, False:

   a) $(x+y)^a = x^a + y^a$

   False.

   b) $x^{a-b+c} = \frac{x^a x^c}{x^b}$

   True.

2) By definition $\pi^{\sqrt{2}} = $ _____.

   $e^{\sqrt{2}\ln \pi}$

3) $\lim\limits_{x \to -\infty} \left(\frac{1}{2}\right)^x = $ _____.

   ∞

4) Graph $y = (0.7)^x$.

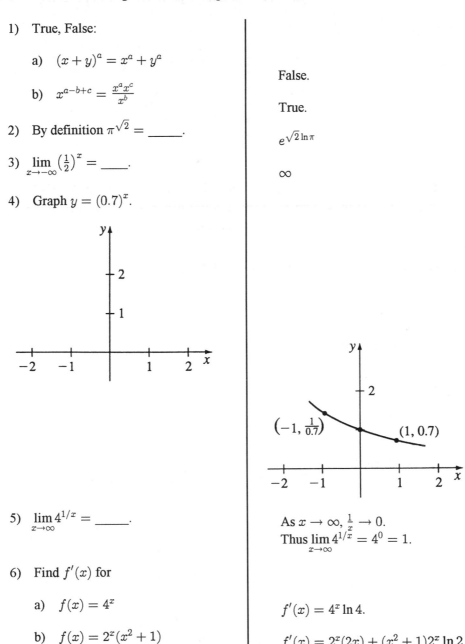

As $x \to \infty$, $\frac{1}{x} \to 0$.
Thus $\lim\limits_{x \to \infty} 4^{1/x} = 4^0 = 1$.

5) $\lim\limits_{x \to \infty} 4^{1/x} = $ _____.

6) Find $f'(x)$ for

   a) $f(x) = 4^x$

   $f'(x) = 4^x \ln 4$.

   b) $f(x) = 2^x(x^2 + 1)$

   $f'(x) = 2^x(2x) + (x^2 + 1)2^x \ln 2$
   $= 2^x(2x + \ln 2(x^2 + 1))$.

7) $\int 7^{-x}\, dx =$

$u = -x,\ du = -dx,\ dx = -du$
$\int 7^{-x}\, dx = \int 7^u(-du)$
$\qquad = -\int 7^u\, du = -\dfrac{7^u}{\ln 7}$
$\qquad = -\dfrac{7^{-x}}{\ln 7} + C.$

8) For what a us $(a^x)' = a^x$?

Only for $a = e$.

B. The general logarithmic function $y = \log_a x$ is defined as the inverse of $y = a^x$. $\log_a x = y$ means $a^y = x$. This definition requires $x > 0$, since $a^y > 0$ for all y. $\log_a x$ is the exponent you put on a to get x.

The function $y = \log_a x$ has these properties:
$\log_a(a^x) = x$
$a^{\log_a x} = x$, for $x > 0$
$\log_a(xy) = \log_a x + \log_a y$
$\log_a\!\left(\dfrac{x}{y}\right) = \log_a x - \log_a y$
$\log_a x^c = c \log_a x$
$\displaystyle\lim_{x\to\infty} \log_a x = \infty$
$\displaystyle\lim_{x\to 0^+} \log_a x = -\infty$

For $a > 1$, the graph of $y = \log_a x$ is

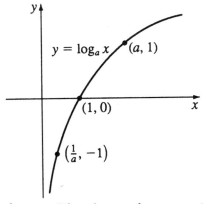

In the special case that $a = e$, $\ln x$ means $\log_a x$. Thus $\ln e = \log_e e = 1$.
$\log_a x = \dfrac{\ln x}{\ln a}$.
If $f(x) = \log_a x$, $f'(x) = \dfrac{1}{x \ln a}$.

9) $\log_4 8 = $ _____.

Let $y = \log_4 8$.
Then $4^y = 8$
$2^{2y} = 2^3$
$2y = 3$
$y = \frac{3}{2}$

10) True or False:

   a) $\log_a x^3 = 3 \log_a x$

True.

   b) $\log_2(10) = \log_2(-5) + \log_2(-2)$

False.

11) For $x > 0$, simplify using properties of $\log_2$: $\log_2\left(\frac{4x^3}{\sqrt{2}}\right)$

$\log_2\left(\frac{4x^3}{\sqrt{2}}\right) = \log_2 4x^3 - \log_2 \sqrt{2}$
$= \log_2 4 + \log_2 x^3 - \log_2 2^{1/2}$
$= \log_2 4 + 3\log_2 x - \frac{1}{2}\log_2 2$
$= 2 + 3\log_2 x - \frac{1}{2}(1)$
$= \frac{3}{2} + 3\log_2 x.$

12) Solve for x:

   a) $e^{x+1} = 10$

By definition, this means $\ln 10 = x + 1$, so $x = \ln 10 - 1$.

   b) $\ln(x + 5) = 2$

By definition, this means $e^2 = x + 5$, so $x = e^2 - 5$.

   c) $2^{3x} = 7$

Take $\log_2$ of both sides:
$\log_2 2^{3x} = \log_2 7$
$3x = \log_2 7$
$x = \frac{1}{3}\log_2 7.$

13) $\log_2 10 = \dfrac{\ln \underline{}}{\ln \underline{}}$

$\log_2 10 = \frac{\ln 10}{\ln 2}$

14) Find $f'(x)$ for $f(x) = \log_2(x^2 + 1)$

$f'(x) = \frac{1}{\ln 2 (x^2+1)} \cdot (2x) = \frac{2x}{\ln 2 (x^2+1)}.$

15) Use logarithmic differentiation to find y' for $y = 6^x$.

$\ln y = \ln 6^x = x \ln 6.$
Thus $\frac{d}{dx} \ln y = \ln 6$ (a constant)
and $y' = y \ln 6 = 6^x \ln 6.$

C. $\lim\limits_{x \to 0^+} (1+x)^{1/x} = e$. Equivalently, $\lim\limits_{t \to \infty} \left(1 + \frac{1}{t}\right)^t = e$.

16) Find each limit.

    a) $\lim\limits_{h \to 0^+} (1+h)^{1/h}$

                              e

    b) $\lim\limits_{x \to \infty} \left(1 + \frac{1}{x}\right)^{3x}$

$$\lim\limits_{x \to \infty} \left(1 + \frac{1}{x}\right)^{3x} = \lim\limits_{x \to \infty} \left[\left(1 + \frac{1}{x}\right)^x\right]^3$$
$$= \left[\lim\limits_{x \to \infty} \left(1 + \frac{1}{x}\right)^x\right]^3 = e^3.$$

    c) $\lim\limits_{x \to 1^+} x^{1/(x-1)}$

Let $t = x - 1$, hence $x = 1 + t$.
As $x \to 1^+$, $t \to 0^+$. Thus
$$\lim\limits_{x \to 1^+} x^{1/(x-1)} = \lim\limits_{t \to 0^+} (1+t)^{1/t} = e.$$

17) Give three different definitions for the number e.

  1. e is the unique number for which $\ln e = 1$.

  2. e is the unique number for which $\lim\limits_{h \to 0^+} \frac{e^h - 1}{h} = 1$.

  3. $e = \lim\limits_{x \to 0^+} (1+x)^{1/x}$.

Exponential Growth and Decay

Concepts to Master

Solution to $y' = ky$; Applications to growth and decay.

Summary and Focus Questions

The solution to the differential equation $\frac{dy}{dt} = ky$ is $y(t) = y(0) \cdot e^{kt}$.

This solution may be used in problems where a quantity (y) of a substance varies with time in such a manner that the rate of change (y') of the quantity is proportional to y, that is, $y' = ky$, for some constant k. If $k > 0$ this is <u>natural growth</u>; if $k < 0$, this is <u>natural decay</u>.

Many growth and decay problems involving $y(t) = y(0)e^{kt}$ boil down to: "given $y = y_1$ at time t_1 and $y = y_2$ at time t_2, find y at some third time t_3." The procedure for solving $y' = ky$ depends on whether or not one of the times given is $t = 0$.

If $t_1 = 0$, then we are given $y(0)$. Then just solve $y_2 = y(0) \cdot e^{kt_2}$ for k.

If neither t_1 nor t_2 are zero, solve the equations $y_1 = y(0) \cdot e^{kt_1}$ and $y_2 = y(0) \cdot e^{kt_2}$ simultaneously for $y(0)$ and k. You first divide one equation by the other to eliminate the $y(0)$ term, and then solve for k.

1) A village had a population of 1000 in 1980 and 1200 in 1990. Assuming the population is experiencing natural growth, what will be the population in 2010?

Let 1980 be $t = 0$. Then $y(0) = 1000$.
At $t_1 = 10$ (year 1990), $y = 1200$
so $1200 = 1000e^{k(10)}$
$1.2 = e^{10k}$
$10k = \ln 1.2$

173

$k = \frac{1}{10} \ln 1.2$

Thus $y = 1000e^{(1/10 \ln 1.2)t}$

$\qquad = 1000(e^{\ln 1.2})^{t/10}$

$\qquad = 1000(1.2)^{t/10}$

In the year 2010, $t = 30$, so

$y = 1000(1.2)^{30/10} = 1000(1.2)^3 \approx 1728.$

2) One hour after a bacteria culture was started there were 300 organisms and after 2 hours from the start there were 900. How many organisms will there be 4 hours after the start?

At $t = 1$, $y = 300$; $t = 2$, $y = 900$.

Thus $300 = y(0)e^{k \cdot 1}$

$900 = y(0)e^{k \cdot 2}$.

Dividing the second equation by the first

$\frac{900}{300} = \frac{y(0)e^{2k}}{y(0)e^k}$; $3 = e^k$; $k = \ln 3.$

Now use $300 = y(0)e^{k \cdot 1}$ to solve for $y(0)$:

$300 = y(0)e^{\ln 3}$; $300 = y(0)3$; $y(0) = 100.$

The model is

$y = 100e^{\ln 3 t} = 100(e^{\ln 3})^t = 100 \; 3^t.$

So $y = 100 \; 3^t.$

At $t = 4$, $y = 100 \; 3^4 = 8100.$

3) A sample of Stronium-90 initially weighs 90 mg. After 100 days 60 mg remain. After how many days will the sample be down to 20 mg in mass?

At $t_1 = 0$, $y(0) = 90.$

At $t_2 = 100$, $y = 60$, so

$60 = 90e^{k(100)}$

$\frac{2}{3} = e^{100k}$; $100k = \ln \frac{2}{3}$; $k = \frac{1}{100} \ln \frac{2}{3}$

Thus $y = 90e^{(1/100 \ln 2/3)t}$

$\qquad = 90(e^{\ln 2/3})^{t/100}$

$\qquad = 90\left(\frac{2}{3}\right)^{t/100}$

When will $y = 20$?

$20 = 90\left(\frac{2}{3}\right)^{t/100}$

$\frac{2}{9} = \left(\frac{2}{3}\right)^{t/100}$

$\ln\left(\frac{2}{9}\right) = \ln\left(\frac{2}{3}\right)^{t/100} = \frac{t}{100} \ln\left(\frac{2}{3}\right)$

$t = \frac{100 \ln \frac{2}{9}}{\ln \frac{2}{3}} \approx 371$ days.

Inverse Trigonometric Functions

Concepts to Master

A. Definitions and functional values of the six inverse trigonometric functions
B. Derivatives and integrals involving inverse trigonometric functions

Summary and Focus Questions

A. For any function f for which f^{-1} is a function, $y = f^{-1}(x)$ if and only if $x = f(y)$.

None of the trigonometric functions is one-to-one. However, when the domains of the trigonometric function are properly restricted they have inverse functions.

For example, sine with domain $\left[-\frac{\pi}{2}, \frac{\pi}{2}\right]$ has an inverse function $\sin^{-1}$. Thus $y = \sin^{-1}\left(\frac{1}{2}\right)$ means $\sin(y) = \frac{1}{2}$ and $-\frac{\pi}{2} \le y \le \frac{\pi}{2}$; so $y = \frac{\pi}{6}$.

The domains and ranges of the six inverse trigonometric functions are:

Function	Domain	Range
$\sin^{-1}$	$[-1, 1]$	$\left[-\frac{\pi}{2}, \frac{\pi}{2}\right]$
$\cos^{-1}$	$[-1, 1]$	$[0, \pi]$
$\tan^{-1}$	all reals	$\left(-\frac{\pi}{2}, \frac{\pi}{2}\right)$
$\cot^{-1}$	all reals	$(0, \pi)$
$\sec^{-1}$	$(-\infty, -1] \cup [1, \infty)$	$\left[0, \frac{\pi}{2}\right) \cup \left[\pi, \frac{3\pi}{2}\right)$
$\csc^{-1}$	$(-\infty, -1] \cup [1, \infty)$	$\left(0, \frac{\pi}{2}\right] \cup \left(\pi, \frac{3\pi}{2}\right]$

The first three functions are used frequently; their graphs are:

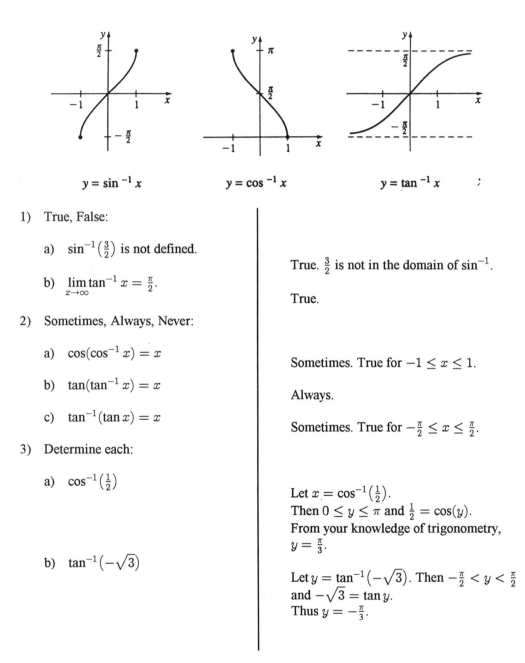

$$y = \sin^{-1} x \qquad\qquad y = \cos^{-1} x \qquad\qquad y = \tan^{-1} x$$

1) True, False:

a) $\sin^{-1}\left(\frac{3}{2}\right)$ is not defined.

True. $\frac{3}{2}$ is not in the domain of $\sin^{-1}$.

b) $\lim\limits_{x \to \infty} \tan^{-1} x = \frac{\pi}{2}$.

True.

2) Sometimes, Always, Never:

a) $\cos(\cos^{-1} x) = x$

Sometimes. True for $-1 \leq x \leq 1$.

b) $\tan(\tan^{-1} x) = x$

Always.

c) $\tan^{-1}(\tan x) = x$

Sometimes. True for $-\frac{\pi}{2} \leq x \leq \frac{\pi}{2}$.

3) Determine each:

a) $\cos^{-1}\left(\frac{1}{2}\right)$

Let $x = \cos^{-1}\left(\frac{1}{2}\right)$.
Then $0 \leq y \leq \pi$ and $\frac{1}{2} = \cos(y)$.
From your knowledge of trigonometry,
$y = \frac{\pi}{3}$.

b) $\tan^{-1}\left(-\sqrt{3}\right)$

Let $y = \tan^{-1}\left(-\sqrt{3}\right)$. Then $-\frac{\pi}{2} < y < \frac{\pi}{2}$
and $-\sqrt{3} = \tan y$.
Thus $y = -\frac{\pi}{3}$.

c) $\sin^{-1}\left(\sin\frac{3\pi}{4}\right)$

Since $\sin\frac{3\pi}{4} = \frac{1}{\sqrt{2}}$, this problem asks you to find $y = \sin^{-1}\left(\frac{1}{\sqrt{2}}\right)$. Then $-\frac{\pi}{2} \le y \le \frac{\pi}{2}$ and $\sin y = \frac{1}{\sqrt{2}}$. Thus $y = \frac{\pi}{4}$.

d) $\sin^{-1}\left(\sin\frac{\pi}{3}\right)$

For $-\frac{\pi}{2} \le x \le \frac{\pi}{2}$, $\sin^{-1}(\sin x) = x$. Thus $\sin^{-1}\left(\sin\frac{\pi}{3}\right) = \frac{\pi}{3}$.

e) $\sec^{-1}\left(\sin\frac{\pi}{7}\right)$

This does not exist because $-1 \le \sin\frac{\pi}{7} \le 1$, so $\sin\frac{\pi}{7}$ is not in the domain of $\sec^{-1}$.

4) For $0 < x < 1$, simplify $\tan(\cos^{-1}x)$.

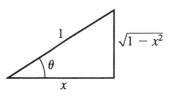

If $\cos\theta = x$ where $0 < \theta < \frac{\pi}{2}$ then $\theta = \cos^{-1}x$ and $\tan\theta = \frac{\sqrt{1-x^2}}{x}$.

B. Here are the derivatives of the inverse trigonometric functions:

$\frac{d}{dx}\sin^{-1}x = \frac{1}{\sqrt{1-x^2}}$ $\qquad\qquad \frac{d}{dx}\cos^{-1}x = \frac{-1}{\sqrt{1-x^2}}$

$\frac{d}{dx}\tan^{-1}x = \frac{1}{1+x^2}$ $\qquad\qquad \frac{d}{dx}\cot^{-1}x = \frac{-1}{1+x^2}$

$\frac{d}{dx}\sec^{-1}x = \frac{1}{x\sqrt{x^2-1}}$ $\qquad\qquad \frac{d}{dx}\csc^{-1}x = \frac{-1}{x\sqrt{x^2-1}}$

Because the derivatives in the right column are the negatives of those in the left column, there are only three associated antiderivatives.

$\int\frac{1}{\sqrt{1-x^2}}\,dx = \sin^{-1}x + C$

$\int\frac{1}{1+x^2}\,dx = \tan^{-1}x + C$

$\int\frac{1}{x\sqrt{x^2-1}}\,dx = \sec^{-1}x + C$

5) Find $f'(x)$ for each:

a) $f(x) = \sin^{-1}(x^2)$

$$f'(x) = \frac{1}{\sqrt{1-(x^2)^2}}(2x) = \frac{2x}{\sqrt{1-x^4}}$$

b) $f(x) = (\tan^{-1} x)^3$

$$f'(x) = 3(\tan^{-1} x)^2 \cdot \frac{1}{1+x^2} = \frac{3(\tan^{-1} x)^2}{1+x^2}$$

c) $f(x) = x\cos^{-1} x$

$$f'(x) = x\left(\frac{-1}{\sqrt{1-x^2}}\right) + \cos^{-1} x(1)$$
$$= \frac{-x}{\sqrt{1-x^2}} + \cos^{-1} x.$$

6) Find y' for $y = \sin^{-1} x + \cos^{-1} x$. What can you conclude about $\sin^{-1} x + \cos^{-1} x$?

$$y' = \frac{1}{\sqrt{1-x^2}} + \frac{-1}{\sqrt{1-x^2}} = 0.$$
Thus y is a constant. To determine the constant, chose $x = 0$:
$$y = \sin^{-1} 0 + \cos^{-1} 0 = 0 + \frac{\pi}{2} = \frac{\pi}{2}.$$
Therefore $\sin^{-1} x + \cos^{-1} x = \frac{\pi}{2}$.

7) Evaluate

a) $\int_1^{\sqrt{3}} \frac{1}{1+x^2}\, dx$

$$\int_1^{\sqrt{3}} \frac{1}{1+x^2}\, dx = \tan^{-1} x \Big|_1^{\sqrt{3}}$$
$$= \tan^{-1}(\sqrt{3}) - \tan^{-1}(1)$$
$$= \frac{\pi}{3} - \frac{\pi}{4} = \frac{\pi}{12}.$$

b) $\int_0^1 \frac{1}{\sqrt{4-x^2}}$

This is almost the form needed for $\sin^{-1} x$ except we have $\sqrt{4-x^2}$ instead of $\sqrt{1-x^2}$. Make the constant 1:
$$\sqrt{4-x^2} = \sqrt{4\left(1-\frac{x^2}{4}\right)} = 2\sqrt{1-\left(\frac{x}{2}\right)^2}.$$

Then
$$\int_0^1 \frac{1}{2\sqrt{1-\left(\frac{x}{2}\right)^2}}\, dx$$
$$= \left(u = \frac{x}{2}\ du = \frac{dx}{2}\right)$$
$$= \int_0^1 \frac{1}{\sqrt{1-u^2}}\, du = \sin^{-1} u \Big|_0^1$$
$$= \sin^{-1}\frac{x}{2} \Big|_0^1 = \sin^{-1}\frac{1}{2} - \sin^{-1} 0$$
$$= \frac{\pi}{6} - 0 = \frac{\pi}{6}.$$

c) $\int \frac{1}{2x\sqrt{x^2-1}}\, dx$

$$\int \frac{1}{2x\sqrt{x^2-1}}\, dx = \frac{1}{2}\int \frac{1}{x\sqrt{x^2-1}}\, dx$$
$$= \frac{1}{2}\sec^{-1} x + C.$$

Hyperbolic Functions

Concepts to Master

A. Define the hyperbolic trigonometric functions
B. Derivatives and integrals involving hyperbolic functions
C. Definition and derivatives of inverse hyperbolic functions

Summary and Focus Questions

A. The hyperbolic sine and cosine functions are:
$$\sinh x = \frac{e^x - e^{-x}}{2} \qquad\qquad \cosh x = \frac{e^x + e^{-x}}{2}$$

The other hyperbolic trigonometric functions are:
$$\tanh x = \frac{\sinh x}{\cosh x} \qquad\qquad \coth x = \frac{\cosh x}{\sinh x}$$

$$\operatorname{sech} x = \frac{1}{\cosh x} \qquad\qquad \operatorname{csch} x = \frac{1}{\sinh x}$$

One reason why they are called hyperbolic is that $\cosh^2 t - \sinh^2 t = 1$ and so $(\cosh t, \sinh t)$ is on the hyperbola $x^2 - y^2 = 1$. (Recall the trigonometric functions are sometimes called "circular" functions because $(\cos t, \sin t)$ is on the circle $x^2 + y^2 = 1$.)

1) Evaluate $\sinh 2$.

2) Write $\coth x$ in terms of e^x.

$$\sinh 2 = \frac{e^2 - e^{-2}}{2} = \frac{e^4 - 1}{2e^2}.$$

$$\coth x = \frac{\cosh x}{\sinh x} = \frac{\frac{e^x + e^{-x}}{2}}{\frac{e^x - e^{-x}}{2}} = \frac{e^x + e^{-x}}{e^x - e^{-x}}.$$

3) Verify $\sinh 2x = 2 \sinh x \cosh x$.

$$2 \sinh x \cosh x = 2 \frac{e^x - e^{-x}}{2} \cdot \frac{e^x + e^{-x}}{2}$$
$$= \frac{(e^x - e^{-x})(e^x + e^{-x})}{2}$$
$$= \frac{e^{2x} - e^{-2x}}{2} = \sinh 2x.$$

B. Here are derivatives and antiderivatives for the hyperbolics:

$$\frac{d}{dx}\sinh x = \cosh x \qquad\qquad \frac{d}{dx}\cosh x = \sinh x$$

$$\frac{d}{dx}\tanh x = \text{sech}^2 x \qquad\qquad \frac{d}{dx}\coth x = -\text{csch}^2 x$$

$$\frac{d}{dx}\text{sech}\, x = -\text{sech}\, x \tanh x \qquad\qquad \frac{d}{dx}\text{csch}\, x = -\text{csch}\, x \coth x$$

$$\int \sinh x\; dx = \cosh x + C \qquad\qquad \int \cosh x\; dx = \sinh x + C$$

$$\int \text{sech}^2 x\; dx = \tanh x + C \qquad\qquad \int \text{csch}^2 x\; dx = -\coth x + C$$

$$\int \text{sech}\, x \tanh x\; dx = -\text{sech}\, x + C \qquad\qquad \int \text{csch}\, x \coth x\; dx = -\text{csch}\, x + C$$

4) Find $f'(x)$ for each:

a) $f(x) = \tanh\left(\frac{x}{2}\right)$

$$f'(x) = \text{sech}^2\left(\frac{x}{2}\right)\left(\frac{1}{2}\right) = \tfrac{1}{2}\,\text{sech}^2\left(\frac{x}{2}\right).$$

b) $f(x) = \sqrt{\cosh x}$

$$f'(x) = \tfrac{1}{2}(\cosh)^{-1/2}(\sinh x) = \frac{\sinh x}{2\sqrt{\cosh x}}$$

c) $f(x) = \sinh x + \cosh x$

$f'(x) = \cosh x + \sinh x$. Thus $f'(x) = f(x)$. If this seems familiar it should since $\sinh x + \cosh x = e^x$ and $(e^x)' = e^x$.

5) Evaluate

a) $\int x \sinh x^2\; dx$

$$u = x^2,\; du = 2x\; dx,\; \frac{du}{2} = x\; dx$$
$$\int x \sinh x^2\; dx = \int \sinh u \frac{du}{2}$$
$$= \tfrac{1}{2}\int \sinh u = \tfrac{1}{2}\cosh u$$
$$= \tfrac{1}{2}\cosh x^2 + C.$$

b) $\int \coth x\; dx$

$$\int \coth x\; dx = \int \frac{\cosh x}{\sinh x}\; dx$$
$$(u = \sinh x,\; du = \cosh x\; dx)$$
$$= \int \frac{1}{u}\; du = \ln|u|$$
$$= \ln|\sinh x| + C.$$

C. The inverses of the hyperbolic functions are:

Function	Domain	Range	Explicit Form
$\sinh^{-1} x$	all reals	all reals	$\ln\left(x + \sqrt{x^2 + 1}\right)$
$\cosh^{-1} x$	$[1, \infty)$	$[0, \infty)$	$\ln\left(x + \sqrt{x^2 - 1}\right)$
$\tanh^{-1} x$	$(-1, 1)$	all reals	$\frac{1}{2}\ln\left(\frac{1+x}{1-x}\right)$
$\coth^{-1} x$	$(-\infty, -1) \cup (1, \infty)$	all reals except 0	$\frac{1}{2}\ln\left(\frac{1+x}{x-1}\right)$
$\text{sech}^{-1} x$	$(0, 1]$	$[0, \infty)$	$\ln\left(\frac{1}{x} + \sqrt{\frac{1}{x^2} - 1}\right)$
$\text{csch}^{-1} x$	all reals except 0	all reals except 0	$\ln\left(\frac{1}{x} + \sqrt{\frac{1}{x^2} + 1}\right)$

The derivatives are:

$$\frac{d}{dx}\sinh^{-1} x = \frac{1}{\sqrt{1+x^2}} \qquad\qquad \frac{d}{dx}\text{csch}^{-1} x = \frac{1}{|x|\sqrt{x^2+1}}$$

$$\frac{d}{dx}\cosh^{-1} x = \frac{1}{\sqrt{x^2-1}} \qquad\qquad \frac{d}{dx}\text{sech}^{-1} x = \frac{-1}{x\sqrt{1-x^2}}$$

$$\frac{d}{dx}\tanh^{-1} x = \frac{1}{1-x^2} \qquad\qquad \frac{d}{dx}\coth^{-1} x = \frac{1}{1-x^2}$$

6) Show that $f(x) = \tanh^{-1} x$ is always increasing.

$f'(x) = \frac{1}{1-x^2}$

Since the domain is $(-1, 1)$ we have $-1 < x < 1$, hence $\frac{1}{1-x^2} > 0$.

7) Verify the derivative $\frac{d}{dx}\coth^{-1} x = \frac{1}{1-x^2}$, using the explicit form for $\coth^{-1}$ x.

$\coth^{-1} x = \frac{1}{2}\ln\left(\frac{1+x}{x-1}\right)$. Therefore

$\frac{d}{dx}\coth^{-1} x = \frac{1}{2}\frac{1}{\frac{1+x}{x-1}}\frac{(x-1)(1)-(1-x)(1)}{(x-1)^2}$

$= \frac{1}{2}\frac{x-1}{1+x}\frac{-2}{(x-1)^2} = \frac{-1}{(1+x)(x-1)}$

$= \frac{-1}{x^2-1} = \frac{1}{1-x^2}$

8) Find $f'(x)$ for:

a) $f(x) = \sinh^{-1} x^2$.

$f'(x) = \frac{1}{\sqrt{1+(x^2)^2}}(2x) = \frac{2x}{\sqrt{1+x^4}}$

b)　$f(x) = \tanh^{-1}\sqrt{1-x^2}$

$$f'(x) = \frac{1}{1-\left(\sqrt{1-x^2}\right)^2}\,\frac{1}{2}(1-x^2)^{-1/2}(-2x)$$
$$= \frac{1}{x^2}\,\frac{1}{2}\,\frac{1}{\sqrt{1-x^2}}(-2x)$$
$$= \frac{-1}{x\sqrt{1-x^2}}$$

Indeterminate Forms and L'Hôpital's Rule

Concepts to Master

Recognition of indeterminate forms; L'Hôpital's Rule

Summary and Focus Questions

Sometimes applying the limit theorems of Chapter 1 results in a nonsensical expression yet the limit still exists. Several of these expressions are called indeterminate forms. For example, $\lim\limits_{x \to 2} \frac{x^2-4}{x-2}$ has the indeterminate form $\frac{0}{0}$ but the limit still exists and is 4.

This section shows you how to deal with seven such forms.

Forms $\frac{0}{0}$ and $\frac{\infty}{\infty}$:

Limits of the form $\lim\limits_{x \to a} \frac{f(x)}{g(x)}$ which have indeterminate form $\frac{0}{0}$ or $\frac{\infty}{\infty}$ are amenable to L'Hôpital's Rule:

$$\lim_{x \to a} \frac{f(x)}{g(x)} = \lim_{x \to a} \frac{f'(x)}{g'(x)}.$$

Example: Although $\lim\limits_{x \to a} \frac{x^2-4}{x-2}$ can be evaluated by factoring, canceling, and applying the limit theorems, L'Hôpital's Rule is easier:

$$\lim_{x \to a} \frac{x^2-4}{x-2} = \lim_{x \to 2} \frac{2x}{1} = 4.$$

Forms $0 \cdot \infty$:

To evaluate $\lim\limits_{x \to a} f(x) \cdot g(x)$ which has indeterminate form $0 \cdot \infty$, first rewrite $f(x) \cdot g(x)$ as either $\frac{f(x)}{1/g(x)}$ or $\frac{g(x)}{1/f(x)}$ to get either $\frac{0}{0}$ or $\frac{\infty}{\infty}$. Now apply L'Hôpital's Rule.

Example: $\lim\limits_{x \to \infty} x e^{-x}$ (form $\infty \cdot 0$) $= \lim\limits_{x \to \infty} \frac{x}{e^x}$ (form $\frac{\infty}{\infty}$) $= \lim\limits_{x \to \infty} \frac{1}{e^x} = 0.$

Form $\infty - \infty$:

For $\lim\limits_{x \to a} f(x) - g(x) = \infty - \infty$, $f(x) - g(x)$ must be rewritten as a single term. When the trigonometric functions are involved, switching to all sines and cosines may help.

<u>Example</u>: $\lim\limits_{x \to \pi/2^-} (\sec x - \tan x) = \lim\limits_{x \to \pi/2^-} \left(\frac{1}{\cos x} - \frac{\sin x}{\cos x}\right) = \lim\limits_{x \to \pi/2^-} \frac{1 - \sin x}{\cos x} \left(\text{form } \frac{0}{0}\right)$

$\qquad\qquad = \lim\limits_{x \to \pi/2^-} \frac{-\cos x}{-\sin x} = 0.$

Forms 0^0, 1^∞, ∞^0:

Finally, for limits of the type $\lim\limits_{x \to a} (f(x))^{g(x)}$, indeterminate forms of type 0^0, 1^∞, and ∞^0 are possible. In these cases,

1. Let $y = f(x)^{g(x)}$.
2. Then $\ln y = g(x) \ln f(x)$.
3. If $\lim\limits_{x \to a} g(x) \ln f(x)$ (form $0 \cdot \infty$) exists and equals L, then
   $$\lim\limits_{x \to a} (f(x))^{g(x)} = e^L.$$

1) True, False:

   All limits having one of the seven indeterminate forms are solved by reducing (if necessary) to limits of the form $\frac{0}{0}$ or $\frac{\infty}{\infty}$.

   True. L'Hôpital's Rule only applies in $\frac{0}{0}$ and $\frac{\infty}{\infty}$ cases. The other five forms reduce to one of these two forms.

2) Evaluate each:

   a) $\lim\limits_{x \to 3} \frac{x^2 - 7x + 12}{x^2 - 9}$

   $\lim\limits_{x \to 3} \frac{x^2 - 7x + 12}{x^2 - 9} \left(\text{form } \frac{0}{0}\right)$
   $\qquad = \lim\limits_{x \to 3} \frac{2x - 7}{2x} = \frac{-1}{6}.$

   b) $\lim\limits_{x \to 0} \frac{\sin x}{x}$

   $\lim\limits_{x \to 0} \frac{\sin x}{x}$ (a familiar limit)
   $\qquad = \lim\limits_{x \to 0} \frac{\cos x}{1} = \frac{1}{1} = 1.$

   c) $\lim\limits_{x \to \infty} \frac{e^x}{3x}$

   $\lim\limits_{x \to \infty} \frac{e^x}{3x} \left(\text{form } \frac{\infty}{\infty}\right) = \lim\limits_{x \to \infty} \frac{e^x}{3} = \infty.$

d) $\lim\limits_{x\to 0^+} x\cot x$

This has form $0\cdot\infty$. Rewrite $x\cot x$ as
$x\cdot\frac{\cos x}{\sin x}=\frac{x}{\sin x}\cdot\cos x$.
$\lim\limits_{x\to 0^+} x\cot x = \lim\limits_{x\to 0^+}\frac{x}{\sin x}\cdot\cos x$

$\qquad = \left(\lim\limits_{x\to 0^+}\frac{x}{\sin x}\right)\cdot\left(\lim\limits_{x\to 0^+}\cos x\right)\left(\text{form } \tfrac{0}{0}\right)$

$\qquad = \lim\limits_{x\to 0^+}\frac{1}{\cos x}\cdot\lim\limits_{x\to 0^+}\cos x = \lim\limits_{x\to 0^+} 1 = 1.$

e) $\lim\limits_{x\to\infty}\frac{1}{x^2-1}-\frac{1}{x-1}$

This limit is $0-0=0$. (Not an indeterminate form.)

f) $\lim\limits_{x\to 0^+}\left(\frac{1}{x}-\frac{1}{e^x-1}\right)$

$\lim\limits_{x\to 0^+}\left(\frac{1}{x}-\frac{1}{e^x-1}\right)(\text{form }\infty-\infty)$

$\qquad = \lim\limits_{x\to 0^+}\frac{e^x-1-x}{x(e^x-1)}\ \left(\text{form }\tfrac{0}{0}\right)$

$\qquad = \lim\limits_{x\to 0^+}\frac{e^x-1}{xe^x+e^x-1}\ \left(\text{form }\tfrac{0}{0}\text{ again}\right)$

$\qquad = \lim\limits_{x\to 0^+}\frac{e^x}{xe^x+2e^x}=\frac{1}{0+2}=\frac{1}{2}.$

g) $\lim\limits_{x\to 0^+}\left(\frac{1}{\sin^2 x}-\frac{\cot x}{x}\right)$

$\lim\limits_{x\to 0^+}\left(\frac{1}{\sin^2 x}-\frac{\cot x}{x}\right)$

$\qquad = \lim\limits_{x\to 0^+}\frac{x-\cos x\sin x}{x\sin^2 x}\ \left(\text{form }\tfrac{0}{0}\right)$

$\qquad = \lim\limits_{x\to 0^+}\frac{1-\cos^2 x+\sin^2 x}{\sin^2 x+2x\sin x\cos x}$

$\qquad = \lim\limits_{x\to 0^+}\frac{2\sin^2 x}{\sin x(\sin x+2x\cos x)}$

$\qquad = \lim\limits_{x\to 0^+}\frac{2\sin x}{\sin x+2x\cos x}\ \left(\text{form }\tfrac{0}{0}\right)$

$\qquad = \lim\limits_{x\to 0^+}\frac{2\cos x}{3\cos x-2x\sin x}=\frac{2}{3}.$

h) $\lim\limits_{x\to 0^+}(\csc x)^x$

This has form ∞^0. Let $y=(\csc x)^x$.
$\ln y = x\ln(\csc x)$.
$\lim\limits_{x\to 0^+} x(\ln\csc x)\ (\text{form }0\cdot\infty)$

$\qquad = \lim\limits_{x\to 0^+}\frac{\ln\csc x}{x^{-1}}\ \left(\text{form }\tfrac{\infty}{\infty}\right)$

$\qquad = \lim\limits_{x\to 0^+}\frac{\frac{1}{\csc x}\cdot(-\csc x\cot x)}{(-1)x^{-2}}$

$\qquad = \lim\limits_{x\to 0^+} x^2\cot x\ (\text{form }0\cdot\infty)$

$\qquad = \lim\limits_{x\to 0^+}\frac{x^2}{\tan x}\ \left(\text{form }\tfrac{0}{0}\right)$

$\qquad = \lim\limits_{x\to 0^+}\frac{2x}{\sec^2 x}=\frac{2(0)}{1}=0.$

Thus $\lim\limits_{x\to 0^+}(\csc x)^x=e^0=1.$

7

Techniques of Integration

"I THINK YOU SHOULD BE MORE EXPLICIT HERE IN STEP TWO."

Cartoons courtesy of Sidney Harris. Used by permission.

Integration by Parts

Concepts to Master

Evaluation of <u>integrals by parts</u>

Summary and Focus Questions

The integration by parts formula
$$\int u \, dv = uv - \int v \, du$$
comes from the product rule for derivatives: $(uv)' = uv' + u'v$.
Integration by parts is just a rule for replacing the problem of evaluating $\int u \, dv$ with the hopefully easier problem of evaluating $\int v \, du$. The key to successful use is to identify the two terms u and v' in $\int u \, dv$ (remember, $u \, dv = uv' \, dv$), and to check that $\int v \, du$ is indeed simplier.

1) Give an appropriate choice for u and dv in each by parts integral and evaluate:

a) $\int x^2 \ln x \, dx$
$u = $ _____
$dv = $ _____

$u = \ln x$, $du = \frac{1}{x} \, dx$.
$dv = x^2 \, dx$, $v = \frac{x^3}{3}$.
Thus $v \, du = \frac{x^3}{3} \cdot \frac{1}{x} \, dx = \frac{x^2}{3} \, dx$ and
$\int x^2 \ln x \, dx = (\ln x) \frac{x^3}{3} - \int \frac{x^2}{3} \, dx$
$\quad = \frac{x^3}{3} \ln x - \frac{1}{3} \int x^2 \, dx$
$\quad = \frac{x^3}{3} \ln x - \frac{1}{9} x^3 + C.$

b) $\int x \cos x \, dx$
$u = $ _____
$dv = $ _____

$u = x$, $du = dx$
$dv = \cos x$, $v = \sin x$.
Then $v \, du = \sin x \, dx$ and
$\int x \cos x \, dx = x \sin x - \int \sin x \, dx$
$\quad = x \sin x - (-\cos x)$
$\quad = x \sin x + \cos x + C.$

2) Evaluate $\int x^2 e^{5x}\, dx$.

Let $u = x^2$ and $dv = e^{5x}\, dx$.
Then $du = 2x\, dx$, $v = \frac{1}{5}e^{5x}$, and
$v\, du = \frac{1}{5}e^{5x}2x\, dx = \frac{2}{5}xe^{5x}\, dx$.

Thus $\int x^2 e^{5x}\, dx = x^2\left(\frac{1}{5}e^{5x}\right) - \int \frac{2}{5}xe^{5x}\, dx$
$= \frac{1}{5}x^2 e^{5x} - \frac{2}{5}\int xe^{5x}\, dx$.

The last integral is easier than the original
so we are on the right track. Apply by parts
to it:

$\int xe^{5x}\, dx = \begin{pmatrix} u=x,\, dv=e^{5x}\, dx \\ du=dx,\, v=\frac{1}{5}e^{5x} \end{pmatrix}$
$= x\left(\frac{1}{5}e^{5x}\right) - \int \frac{1}{5}e^{5x}\, dx$
$= \frac{1}{5}xe^{5x} - \frac{1}{25}e^{5x}$.

Thus the original integral is
$\frac{1}{5}x^2 e^{5x} - \frac{2}{5}\left(\frac{1}{5}xe^{5x} - \frac{1}{25}e^{5x}\right)$

$= \frac{1}{5}x^2 e^{5x} - \frac{2}{25}xe^{5x} + \frac{2}{125}e^{5x} + C$.

3) Evaluate $\int \sin x \cos x\, dx$ by parts.

$\int \sin x \cos x\, dx$
$\begin{pmatrix} u = \sin x & dv = \cos x\, dx \\ du = \cos x\, dx & v = \sin x \end{pmatrix}$
$= (\sin x)(\sin x) - \int (\sin x)\cos x\, dx$.

Thus $\int \sin x \cos x\, dx$
$= \sin^2 x - \int \sin x \cos x\, dx$.

Hence $2\int \sin x \cos x\, dx = \sin^2 x$,
so $\int \sin x \cos x\, dx = \frac{1}{2}\sin^2 x + C$.

Note: This integral could have been
evaluated with the simple substitution
$u = \sin x$, $du = \cos x\, dx$ to yield $\int u\, du$.
We could also have solved this with the
identity $\sin x \cos x = \frac{1}{2}\sin 2x$ and the
substitution $u = 2x$. We chose the above
method to illustrate integration by parts.

Trigonometric Integrals

Concepts to Master

Evaluation of integrals using trigonometric identities

Summary and Focus Questions

In this section integrands with certain combinations of trigonometric functions are rewritten using trigonometric identities. The resulting integrals will be easier to evaluate. A summary of the techniques is given in the table below. Note that n and m must be nonnegative integers. If the condition(s) hold, the original integral may be rewritten as indicated.

Original Integral	**Condition(s)**	**Identity to Use**	**Rewritten Form**
	$m = $ odd	$\sin^2 x = 1 - \cos^2 x$	$\int (1 - \cos^2 x)^{(m-1)/2} \cos^n x \sin x \, dx$
$\int \sin^m x \cos^n x \, dx$	$n = $ odd	$\cos^2 x = 1 - \sin^2 x$	$\int (1 - \sin^2 x)^{(n-1)/2} \sin^m x \cos x \, dx$
	m, n both	$\sin^2 x = \frac{1 - \cos 2x}{2}$,	$\int \left(\frac{1 - \cos 2x}{2} \right)^{m/2} \left(\frac{1 + \cos 2x}{2} \right)^{n/2} dx$
	even	$\cos^2 x = \frac{1 + \cos 2x}{2}$	
	$m = $ odd	$\tan^2 x = \sec^2 x - 1$	$\int (\sec^2 x - 1)^{(m-1)/2} \sec^{n-1} x \tan x \sec x \, dx$
$\int \tan^m x \sec^n x \, dx$	$n = $ even	$\sec^2 x = \tan^2 x + 1$	$\int \tan^m x (\tan^2 x + 1)^{(n-2)/2} \sec^2 x \, dx$
	$m = $ even, $n = $ odd	$\tan^2 x = \sec^2 x - 1$	$\int (\sec^2 x - 1)^{m/2} \sec^n x \, dx$
$\int \sin mx \sin nx \, dx$		$\sin A \sin B$ $= \frac{\cos(A-B) - \cos(A+B)}{2}$	$\int \frac{1}{2}(\cos(m-n)x - \cos(m+n)x) dx$
$\int \cos mx \cos nx \, dx$		$\cos A \cos B$ $= \frac{\cos(A-B) + \cos(A+B)}{2}$	$\int \frac{1}{2}(\cos(m-n)x + \cos(m+n)x) dx$
$\int \sin mx \cos nx \, dx$		$\sin A \cos B$ $= \frac{\sin(A-B) + \sin(A+B)}{2}$	$\int \frac{1}{2}(\sin(m+n)x + \sin(m-n)x) dx$

Integrals containing $\cot^m x \csc^n x$ are handled similarly to $\int \tan^m x \sec^n x \, dx$ using the identity $\cot^2 x = \csc^2 x - 1$.

1) Evaluate $\int \sin^5 x \; dx$.

This fits the form $\int \sin^m x \cos^n x \; dx$ with $m = 5$, $n = 0$.
$$\int \sin^5 x \; dx = \int \sin^4 x \sin x \; dx$$
$$= \int (1 - \cos^2 x)^2 \sin x \; dx$$
$$(u = \cos x, \; du = -\sin x \; dx)$$
$$= -\int (1 - u^2)^2 \; du$$
$$= -\int (1 - 2u^2 + u^4) du$$
$$= -u + \tfrac{2}{3} u^3 - \tfrac{1}{5} u^5$$
$$= -\cos x + \tfrac{2}{3} \cos^3 x - \tfrac{1}{5} \cos^5 x + C.$$

2) Evaluate $\int \tan x \sec^4 x \; dx$.

Here $m = 1$ and $n = 4$ so we have our choice of identities. We select the $n = $ even case.
$$\int \tan x \sec^4 x \; dx$$
$$= \int \tan x (\sec^2 x) \sec^2 x \; dx$$
$$= \int \tan x (\tan^2 x + 1) \sec^2 x \; dx$$
$$= \int (\tan^3 x + \tan x) \sec^2 x \; dx$$
$$\left(\begin{matrix} u = \tan x \\ du = \sec^2 x \; dx \end{matrix} \right)$$
$$= \int (u^3 + u) du = \tfrac{u^4}{4} + \tfrac{u^2}{2}$$
$$= \tfrac{1}{4} \tan^4 x + \tfrac{1}{2} \tan^2 x + C.$$

3) Evaluate $\int \sin 6x \cos 2x \; dx$.

The form is $\sin A \cos B$ where $A = 6x$ and $B = 2x$.
$$\int \sin 6x \cos 2x \; dx$$
$$= \int \tfrac{1}{2} (\sin 4x + \sin 8x) dx$$
$$= \tfrac{1}{2} \int \sin 4x \; dx + \tfrac{1}{2} \int \sin 8x \; dx$$
$$\text{(Two substitutions:}$$
$$u = 4x \text{ and } u = 8x.)$$
$$= \tfrac{1}{2} \left(\tfrac{-\cos 4x}{4} \right) + \tfrac{1}{2} \left(\tfrac{-\cos 8x}{8} \right)$$
$$= -\tfrac{1}{8} \cos 4x - \tfrac{1}{16} \cos 8x + C.$$

4) Evaluate $\int \cos^4 x \, dx$.

Here $m = 0$ and $n = 4$ are both even.
$\int \cos^4 x \, dx = \int (\cos^2 x)^2 \, dx$

$= \int \left(\frac{1+\cos 2x}{2} \right)^2 \, dx$

$= \frac{1}{4} \int (1 + 2\cos 2x + \cos^2 2x) dx$

$= \frac{1}{4} \int \left(1 + 2\cos 2x + \frac{1+\cos 4x}{2} \right) dx$

$= \frac{1}{4} \int \left(\frac{3}{2} + 2\cos 2x + \frac{1}{2} \cos 4x \right) dx$

$= \frac{1}{4} \left[\frac{3}{2}x + \sin 2x + \frac{1}{2} \frac{\sin 4x}{4} \right]$

$= \frac{3}{8}x + \frac{1}{4} \sin 2x + \frac{1}{32} \sin 4x + C.$

5) Evaluate $\int \tan^2 x \sec x \, dx$.

Here $m = 2$ is even and $n = 1$ is odd.
$\int \tan^2 x \sec x \, dx = \int (\sec^2 x - 1) \sec x \, dx$
$= \int (\sec^3 x - \sec x) dx$
$= \int \sec^3 x \, dx - \int \sec x \, dx$
$= \frac{1}{2} (\sec x \tan x + \ln |\sec x + \tan x|)$
$\qquad - \ln |\sec x + \tan x| + C$
$= \frac{1}{2} (\sec x \tan x - \ln |\sec x + \tan x|) + C.$
($\int \sec^3 x \, dx$ is evaluated by parts)

Trigonometric Substitution

Concepts to Master

Inverse substitution, trigonometric substitution for integrals containing powers of $\sqrt{a^2 - x^2}$, $\sqrt{a^2 + x^2}$, $\sqrt{x^2 - a^2}$

Summary and Focus Questions

The process of <u>inverse substitution</u> for evaluating $\int f(x) \, dx$ replaces x (and dx) by a function of another variable: t, θ, ... The result, possibly after some simplification, may be an integral easier to evaluate.

This section uses inverse substitutions of the form $x = a \sin \theta$, $x = a \tan \theta$, and $x = a \sec \theta$. The resulting integrals are then evaluated by the techniques of Section 7.2. The table below summarizes the substitutions.

A mnemonic triangle may be drawn to help determine the correct trigonometric substitution. Remember that the triangle is drawn for $0 \leq \theta \leq \frac{\pi}{2}$ only, relationships should be worked out using the identities $\sin^2 \theta + \cos^2 \theta = 1$ and $\tan^2 \theta = \sec^2 \theta - 1$.

Integrand contains	Mnemonic triangle	Substitution
$\sqrt{a^2 - x^2}$		$x = a \sin \theta$, where $-\frac{\pi}{2} \leq \theta \leq \frac{\pi}{2}$ $dx = a \cos \theta \, d\theta$ $\sqrt{a^2 - x^2} = a \cos \theta$.
$\sqrt{a^2 + x^2}$		$x = a \tan \theta$, where $-\frac{\pi}{2} \leq \theta \leq \frac{\pi}{2}$ $dx = a \sec^2 \theta \, d\theta$ $\sqrt{a^2 + x^2} = a \sec \theta$.

Integrand contains	**Mnemonic triangle**	**Substitution**
$\sqrt{x^2 - a^2}$		$x = a\sec\theta$, where $0 < \theta < \frac{\pi}{2}$ or

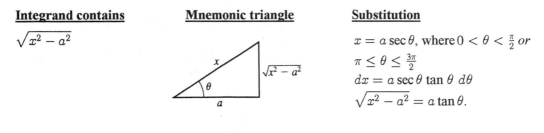

$x = a\sec\theta$, where $0 < \theta < \frac{\pi}{2}$ or
$\pi \le \theta \le \frac{3\pi}{2}$
$dx = a\sec\theta\tan\theta\,d\theta$
$\sqrt{x^2 - a^2} = a\tan\theta.$

It may be necessary to algebraically transform an integrand before substitution. One technique is completing the square:

<u>Example</u>: Evaluate $\int \frac{dx}{\sqrt{x^2+4x+5}}$.

The integrand $\frac{1}{\sqrt{x^2+4x+5}}$ does not apparently meet the form $\frac{1}{\sqrt{1+u^2}}$, but will do so after completing the square:
$$x^2 + 4x + 5 = x^2 + 4x + 4 + 5 - 4 = (x+2)^2 + 1.$$
Thus $\int \frac{dx}{\sqrt{x^2+4x+5}} = \int \frac{dx}{\sqrt{1+(x+2)^2}}$.

Let $u = x + 2$. Then $du = dx$ and $\int \frac{dx}{\sqrt{1+(x+2)^2}} = \int \frac{1}{\sqrt{1+u^2}}\,du.$

We now use a trigonometric substitution:

$\frac{u}{1} = \tan\theta,$

$\frac{\sqrt{1+u^2}}{1} = \sec\theta$

Let $u = \tan\theta$, $du = \sec^2\theta\,d\theta$, $\sqrt{1+u^2} = \sec\theta$.
Then $\int \frac{1}{\sqrt{1+u^2}}\,du = \int \frac{1}{\sec\theta}\sec^2\theta\,d\theta = \int \sec\theta\,d\theta$
$$= \ln\left|\sec\theta + \tan\theta\right| = \ln\left|\sqrt{1+u^2} + u\right|$$
$$= \ln\left|\sqrt{x^2+4x+5} + x + 2\right| + C.$$

1) Rewrite each of the following using a trigonometric substitution:

a) $\int x^2\sqrt{4-x^2}\,dx$

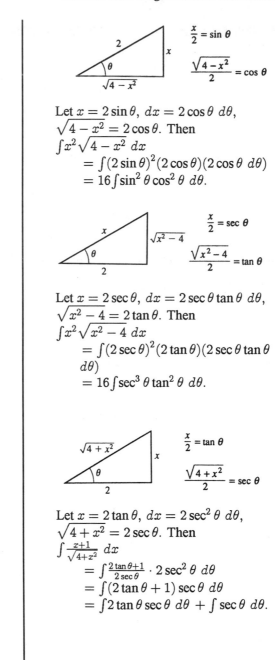

b) $\int x^2 \sqrt{x^2 - 4}\ dx$

$\dfrac{x}{2} = \sec \theta$

$\dfrac{\sqrt{x^2 - 4}}{2} = \tan \theta$

Let $x = 2 \sec \theta$, $dx = 2 \sec \theta \tan \theta\ d\theta$,
$\sqrt{x^2 - 4} = 2 \tan \theta$. Then
$\int x^2 \sqrt{x^2 - 4}\ dx$
$\qquad = \int (2 \sec \theta)^2 (2 \tan \theta)(2 \sec \theta \tan \theta\ d\theta)$
$\qquad = 16 \int \sec^3 \theta \tan^2 \theta\ d\theta.$

c) $\int \dfrac{x+1}{\sqrt{4+x^2}}\ dx$

$\dfrac{x}{2} = \tan \theta$

$\dfrac{\sqrt{4+x^2}}{2} = \sec \theta$

Let $x = 2 \tan \theta$, $dx = 2 \sec^2 \theta\ d\theta$,
$\sqrt{4 + x^2} = 2 \sec \theta$. Then
$\int \dfrac{x+1}{\sqrt{4+x^2}}\ dx$
$\qquad = \int \dfrac{2 \tan \theta + 1}{2 \sec \theta} \cdot 2 \sec^2 \theta\ d\theta$
$\qquad = \int (2 \tan \theta + 1) \sec \theta\ d\theta$
$\qquad = \int 2 \tan \theta \sec \theta\ d\theta + \int \sec \theta\ d\theta.$

3) Rewrite the following so that a trigonometric substitution may be used.
$\int \dfrac{x^2}{\sqrt{13 - 6x + x^2}}\ dx$

$\sqrt{13 - 6x + x^2}$ does not look friendly
until we complete the square:
$$13 - 6x + x^2 = 13 + (x^2 - 6x + 9 - 9)$$
$$= x^2 - 6x + 9 + 4 = (x - 3)^2 + 4.$$
Let $u = x - 3$, $du = dx$, $x = u + 3$, and
$13 - 6x + x^2 = u^2 + 4$.
Then $\int \frac{x^2}{\sqrt{13 - 6x + x^2}} \, dx = \int \frac{(u+3)^2}{\sqrt{u^2 + 4}} \, du.$
The integral is now ready for the
substitution $u = 2\tan\theta$, du$= 2\sec^2\theta \, d\theta$.

Integration of Rational Functions
by Partial Fractions

Concepts to Master

Partial fraction form; Determining term coefficients

Summary and Focus Questions

A proper rational function is a function of the form $\frac{P(x)}{Q(x)}$ where $P(x)$ and $Q(x)$ are polynomials with the degree of $P(x)$ less than the degree of $Q(x)$. Any rational function may be written, using long division if necessary, as a polynomial plus a proper rational function.

The <u>method of partial fractions</u> for evaluating integrals is algebra: write a proper $\frac{P(x)}{Q(x)}$ integrand as a sum of rational functions each of which you already know how to integrate.

Step 1. Write $Q(x)$ in factored form as a product of powers of distinct linear and irreducible quadratics.

Step 2. Write $\frac{P(x)}{Q(x)} = T_1 + T_2 + \ldots + T_n$

where the T_i are terms corresponding to the distinct factors of $Q(x)$ found in Step 1. The forms of T_i are given in the table below.

Type of $Q(x)$ factor	Corresponding T_i term
$x - a$	$\frac{A}{x-a}$
$(x - a)^k$	$\frac{A_1}{x-a} + \frac{A_2}{(x-a)^2} + \ldots + \frac{A_k}{(x-a)^k}$
$ax^2 + bx + c$	$\frac{Ax+B}{ax^2+bx+c}$
$(ax^2 + bx + c)^k$	$\frac{A_1 x+B_1}{ax^2+bx+c} + \frac{A_2 x+B_2}{(ax^2+bx+c)^2} + \ldots + \frac{A_k x+B_k}{(ax^2+bx+c)^k}$

Step 3. Write $P(x) = Q(x)[T_1 + T_2 + \ldots + T_n]$. Then multiply out the right side, combining like terms.

Step 4. Equate coefficients of like terms in the equation in Step 3. This gives a system of linear equations.

Step 5. Solve the linear system in Step 4. This is frequently done by solving for one variable in one equation and using the result to eliminate that variable in the other equations.

Shortcut: Sometimes some of the unknown coefficients can be found after Step 2 by judicious substitution of x values. The rest of the coefficients may be found by continuing with Steps 3, 4, and 5.

Here is a complete example: $\int \frac{4x^3 - 4x^2 + 5x + 1}{(x^2 - 2x + 1)(x^2 + x + 1)}\, dx$.

The integrand is proper because the degree of the numerator is less than the degree of the denominator.

Step 1: $Q(x) = (x^2 - 2x + 1)(x^2 + x + 1) = (x - 1)^2(x^2 + x + 1)$.
      Note that $x^2 + x + 1$ is irreducible.

Step 2: $\frac{4x^3 - 4x^2 + 5x + 1}{(x-1)^2(x^2 + x + 1)} = \frac{A}{(x-1)} + \frac{B}{(x-1)^2} + \frac{Cx + D}{x^2 + x + 1}$.

Step 3: Multiply by the denominator $Q(x)$:
$$4x^3 - 4x^2 + 5x + 1$$
$$= A(x - 1)(x^2 + x + 1) + B(x^2 + x + 1) + (Cx + D)(x - 1)^2.$$
Multiply out the right side:
$$4x^3 - 4x^2 + 5x + 1$$
$$= Ax^3 - A + Bx^2 + Bx + B + Cx^3 - 2Cx^2 + Cx$$
$$+ Dx^2 - 2Dx + D$$
and collect like terms:
$$4x^3 - 4x^2 + 5x + 1$$
$$= (A + C)x^3 + (B - 2C + D)x^2 + (B - 2C + D)x^2$$
$$+ (B + C - 2D)x + (-A + B + D).$$

Step 4: Equate coefficients from the equality in Step 3:
$$4 = A + C$$
$$-4 = B - 2C + D$$
$$5 = B + C - 2D$$
$$1 = -A + B + D$$

Step 5: Solve the system.
From the first equation $A = 4 - C$ and thus the others become:

$-4 = B - 2C + D$	$-4 = B - 2C + D$
$5 = B + C - 2D$ or	$5 = B + C - 2D$
$1 = -(4 - C) + B + D$	$5 = B + C + D$

Subtracting the last two equations yields $0 = 3D$ or $D = 0$.
So the system is:
$$-4 = B - 2C$$
$$5 = B + C.$$

Again subtracting, $-9 = -3C$, $C = 3$. Thus $B = 2$ and $A = 1$.

Therefore $\frac{4x^3 - 4x^2 + 5x + 1}{(x^2 - 2x + 1)(x^2 + x + 1)} = \frac{1}{x-1} + \frac{2}{(x-1)^2} + \frac{3x}{x^2 + x + 1}$.

Finally, the original integral may be written as the sum of three integrals, each of which we know how to evaluate:

$$\int \frac{4x^3 - 4x^2 + 5x + 1}{(x^2 - 2x + 1)(x^2 + x + 1)} \ dx = \int \frac{1}{x-1} \ dx + \int \frac{2}{(x-1)^2} \ dx + \int \frac{3x}{x^2 + x + 1} \ dx \qquad \text{(Whew!)}$$

(In this example we could have found $B = 2$ by setting $x = 1$ in Step 3. The rest of the procedure would be followed to find A, C, and D.)

1) Write the partial fraction form
   (Step 2) for each.

   a) $\frac{2x+1}{x(x+8)}$

   $\frac{A}{x} + \frac{B}{x+8}$

   b) $\frac{4x^2 + 11}{(x+1)^2(x^2 - 3x + 5)}$

   $\frac{A}{x+1} + \frac{B}{(x+1)^2} + \frac{Cx+D}{x^2 - 3x + 5}$

c) $\dfrac{3}{x(x^2+2x+2)^2}$

d) $\dfrac{x^3}{x^2+2x+1}$

2) Evaluate $\int \dfrac{6-x}{x(x+3)}\, dx$.

$\dfrac{A}{x} + \dfrac{Bx+C}{x^2+2x+2} + \dfrac{Dx+E}{(x^2+2x+2)^2}$

First write the fraction (using long division):
$$x - 2 + \frac{3x+2}{x^2+2x+1} = x - 2 + \frac{3x+2}{(x+1)^2}.$$
The partial fraction form is
$$x - 2 + \frac{A}{x+1} + \frac{B}{(x+1)^2}.$$

(1) $Q(x) = x(x+3)$

(2) $\dfrac{6-x}{x(x+3)} = \dfrac{A}{x} + \dfrac{B}{x+3}$

(3) Clear fractions:
$$6 - x = A(x+3) + Bx$$
Combine terms:
$$6 - x = (A+B)x + 3A$$

(4) Equate coefficients:
$$A + B = -1$$
$$3A = 6$$

(5) Solving, we see $A = 2$ from the second equation. Thus
$2 + B = -1$, $B = -3$.
Therefore
$$\int \frac{6-x}{x(x+3)}\, dx = \int\left(\frac{2}{x} + \frac{-3}{x+3}\right) dx$$
$$= 2\ln|x| - 3\ln|x+3| + C.$$

Note: At Step 3, after clearing fractions to get $6 - x = A(x+3) + Bx$ we could substitute $x = 0$ to get $6 = A(0+3) + 0$, so $A = 2$, and we could substitute $x = -3$ to get $6 - (-3) = A(0) + B(-3)$
$9 = -3B$, $B = -3$.
<u>Sometimes</u>, substituting carefully selected values of x can shorten the work.

Rationalizing Substitutions

Concepts to Master

A. Substitution of $u = \sqrt[n]{g(x)}$

B. Substitution of $t = \tan \frac{x}{2}$

Summary and Focus Questions

A. In some cases the substitution $u = \sqrt[n]{g(x)}$ in the form $u^n = g(x)$ will transform an integrand involving radicals into a rational function. Such is the case with $\int \frac{x}{\sqrt{x-2}} \, dx$.

Let $u = \sqrt{x - 2}$, $u^2 = x - 2$, $x = u^2 + 2$, $dx = 2u \, du$.
Then $\int \frac{x}{\sqrt{x-2}} \, dx = \int \frac{u^2+2}{u} 2u \, du = 2 \int (u^2 + 2) du = \frac{2u^3}{3} + 4u$
$\qquad\qquad = \frac{2}{3}(x - 2)^{3/2} + 4\sqrt{x - 2} + C.$

1) Evaluate $\int \frac{\sqrt[3]{x}}{\sqrt[3]{x}+1} \, dx$.

Let $u = \sqrt[3]{x}$, $x = u^3$, $dx = 3u^2 \, du$.
The integral becomes
$\int \frac{u}{u+1} 3u^2 \, du = 3 \int \frac{u^3}{u+1} \, du$.

By long division $\frac{u^3}{u+1} = u^2 - u + 1 + \frac{-1}{u+1}$.
$\int \frac{u^3}{u+1} \, du = \int \left(u^2 - u + 1 - \frac{1}{u+1} \right) du$
$\qquad = \frac{u^3}{3} - \frac{u^2}{2} + u - \ln|u + 1|$.

Since $u = \sqrt[3]{x}$, this is
$3 \left[\frac{x}{3} - \frac{x^{2/3}}{2} + \sqrt[3]{x} - \ln|\sqrt[3]{x} + 1| \right] + C.$

2) What substitution should be made for $\int \frac{\sqrt[3]{x}}{1+\sqrt[4]{x}} \, dx$?

$u = \sqrt[12]{x}$, because then $\sqrt[4]{x}(= u^3)$ and $\sqrt[3]{x}(= u^4)$ are both integer powers of u. (*Note*: 12 is the least common multiple of 3 and 4.)

B. Rational functions of $\sin x$ and $\cos x$ can be integrated using the substitution $t = \tan \frac{x}{2}$. This results in

$$\sin x = \tfrac{2t}{1+t^2}, \ \cos x = \tfrac{1-t^2}{1+t^2}, \ dx = \tfrac{2}{1+t^2} \ dt.$$

Substituting these into the original integrand yields a rational function in the variable t.

3) Evaluate, using the substitution $t = \tan \frac{x}{2}$:

$\int \frac{\sin x}{\cos x + 1} \ dx.$

We make the substitutions:

$\int \frac{\sin x}{\cos x + 1} \ dx = \int \frac{\frac{2t}{1+t^2}}{\frac{1-t^2}{1+t^2}+1} \frac{2}{1+t^2} \ dt$

$= \int \frac{2t}{1-t^2+1+t^2} \frac{2}{1+t^2} \ dt$

$= \int \frac{2}{1+t^2} \ dt = \ln|1 + t^2|$

$= \ln\left(1 + \tan^2 \frac{x}{2}\right) + C.$

4) Write the integral of the rational function resulting from the substitution $t = \tan \frac{x}{2}$ in $\int (1 + \sin^2 x - \cos^2 x) dx.$

The substitution yields

$\int \left[1 + \left(\tfrac{2t}{1+t^2}\right)^2 - \left(\tfrac{1-t^2}{1+t^2}\right)^2 \right] \tfrac{2}{1+t^2} \ dt$

which after some algebraic adjustments becomes $\int \frac{16t^2}{(1+t^2)^3} \ dt$. This may be evaluated using the method of partial fractions.

Strategy for Integration

Concepts to Master

A. No new concepts about integration; just putting together all the integration techniques

B. Functions that are not readily integrable

Summary and Focus Questions

A. The two main techniques for evaluating integrals are: <u>substitution</u> and <u>integration by parts</u>. Substitution can take many forms:

Method	Sample
Straight substitution	$\int \frac{x^2}{x^3+1}\, dx,\ u = x^3 + 1,\ \ldots$
Integration by parts	$\int x^3 \ln x\, dx,\ u = \ln x,\ du = x^3\, dx,\ \ldots$
Trigonometric	$\int \sin^3 x \cos^2 x\, dx = \int \sin^2 x \cos^2 x \sin x\, dx$
	$\quad = \int (1 - \cos^2 x) \cos^2 x \sin x\, dx,\ u = \cos x \ldots$
Trigonometric substitution	$\int \frac{x^2}{\sqrt{4-x^2}}\, dx,\ x = 2 \sin\theta,\ \ldots$
Partial fractions	$\int \frac{1}{(x+2)(x-3)}\, dx = \int \left(\frac{-1}{x+2} + \frac{1}{x-3}\right) dx = \ldots$
Rationalizing substitution	$\int \frac{x}{\sqrt{x+1}}\, dx,\ u = \sqrt{x},\ u^2 = x,\ \ldots$

Remember your algebra and trigonometric identities and know the integrals in the Table of Integration formulas (7.27) of the text.

1) Indicate what technique should be used to begin solving each:

   a) $\int \frac{e^x}{e^x - 1}\, dx$

                           Straight substitution, $u = e^x - 1$.

   b) $\int x^2 \sqrt{16 + x^2}\, dx$

                           Trigonometric substitution, $x = 4 \tan\theta$.

c) $\int \frac{2x+1}{x^3(x+1)} \, dx$

Partial fractions, $\frac{A}{x} + \frac{B}{x^2} + \frac{C}{x^3} + \frac{D}{x+1}$.

d) $\int \tan^3 x \sec^4 x \, dx$

Trigonometric integral,
$\int \tan^3 x \sec^2 x (\sec^2 x \, dx)$.

e) $\int x \sec^2 x \, dx$

Integration by parts,
$u = x$, $dv = \sec^2 x \, dx$.

f) $\int \frac{\csc^2 x}{1+\cot^2 x} \, dx$

Since $1 + \cot^2 x = \csc^2 x$, this one reduces to $\int 1 \, dx = x + C$.

B. All the functions you have encountered so far are "elementary", not because they are considered easy, but because it is possible to express them directly as combinations (addition, subtraction, multiplication, division, composition) of polynomial, rational, exponential, logarithmic, trigonometric, inverse trigonometric, hyperbolic, and inverse hyperbolic functions.

The *derivative* of an elementary function is elementary.
The *antiderivative* of an elementary function need not be elementary.
Thus there are some elementary functions for which there is no simple form for its antiderivative. $\int \frac{1}{\ln x} \, dx$ is one; we will wait until Chapter 10 for this one.

2) True or False:
$f(x) = e^{x^x}$ is elementary.

True.

3) True or False:
If $f(x)$ is not elementary, then $\int f(x) \, dx$ is not elementary.

True.

4) True or False:
If $f(x)$ is elementary, then $\int f(x) \, dx$ is elementary.

False.

5) Do any of the techniques we have solve $\int \cos x^2 \, dx$?

Nope. This one also must wait until Chapter 10.

Approximate Integration

Concepts to Master

A. Approximation of definite integrals using the Midpoint, Trapezoidal, and Simpson's Rule

B. Maximum error estimation

Summary and Focus Questions

A. Suppose f is an integrable function on a closed interval $[a, b]$ which has been partitioned by $\{x_0, x_1, x_2, \ldots, x_n\}$ into n equal intervals of length $\Delta x = \frac{b-a}{n}$. Here are three numbers, M_n, T_n, and S_n, that approximate $\int_a^b f(x)\, dx$.

By the <u>Midpoint Rule</u>, if $\overline{x}_i = \frac{x_{i-1}+x_i}{2}$ (the midpoint of the $[x_{i-1}, x_i]$), then

$$M_n = \Delta x[f(\overline{x_1}) + f(\overline{x_2}) + \ldots + f(\overline{x_n})].$$

M_n is a Riemann sum using the midpoint of each subinterval.

By the <u>Trapezoidal Rule</u>,

$$T_n = \frac{\Delta x}{2}[f(x_0) + 2f(x_1) + 2f(x_2) + \ldots + 2f(x_{n-1}) + f(x_n)].$$

This approximation is derived by computing the sums of areas of inscribed trapezoids.

By <u>Simpson's Rule</u>, if n is even,

$$S_n = \frac{\Delta x}{3}[f(x_0) + 4f(x_1) + 2f(x_2) + 4f(x_3) + \ldots + 4f(x_{n-1}) + f(x_n)].$$

This approximation is derived by replacing portions of the function $y = f(x)$ with approximating parabolas and summing the areas under these parabolas.

1) Estimate $\int_{-1}^{5} 2^x \, dx$ to three decimal places with $n = 6$ subintervals using:

   a) Midpoint Rule.

$\Delta x = \frac{5-(-1)}{6} = 1.$
$x_0 = -1, \, x_1 = 0, \, x_2 = 1, \, \ldots, \, x_6 = 5.$
Thus $\overline{x_1} = -0.5, \, \overline{x_2} = 0.5, \, \overline{x_3} = 1.5,$
$\ldots, \, \overline{x_6} = 4.5.$
$M_n = 1[2^{-0.5} + 2^{0.5} + 2^{1.5} + \ldots + 2^{4.5}]$
$\qquad = 2^{-0.5}[1 + 2 + 2^2 + 2^3 + 2^4 + 2^5]$
$\qquad = \frac{63}{\sqrt{2}} \approx 44.547.$

   b) Trapezoidal Rule

$T_n = \frac{1}{2}[2^{-1} + 2(2^0) + 2(2) + 2(2^2)$
$\qquad + 2(2^3) + 2(2^4) + 2^5]$
$\qquad = \frac{1}{2}(94.5) = 47.250.$

   c) Simpson's Rule

$S_n = \frac{1}{3}[2^{-1} + 4(2^0) + 2(2) + 4(2^2)$
$\qquad + 2(2^3) + 4(2^4) + 2^5]$
$\qquad = \frac{1}{3}(136.5) = 45.500.$

$\left(\text{The actual value of } \int_1^5 2^x \, dx \text{ is } \frac{63}{\ln 4} \approx 45.445.\right)$

2) For $\int_1^5 (40 - x)dx$ and any n, the Midpoint approximation will be:
   a) too large
   b) too small
   c) exact

c) exact, because $y = 40 - x$ is linear. The area of the midpoint rectangle will equal the area under $y = 40 - x$.

3) For $\int_1^5 (40 - x^2)dx$ and any n, the trapezoidal approximation will be:
   a) too large
   b) too small
   c) exact

b) too small. Since the graph of $y = 40 - x^2$ is concave downward, all inscribed trapezoids will have area less than the area under $y = 40 - x^2$.

4) For $\int_1^5 (40 - x^2)dx$ and any even n, Simpson's Rule approximation will be:
   a) too large
   b) too small
   c) exact

c) exact, because Simpson's Rule replaces the function with portions of approximating parabolas and since $y = 40 - x^2$ is a parabola, the approximation is exact.

B. The <u>error of estimation</u> is the difference between the actual value of $\int_a^b f(x)dx$ and the estimate. Here is a table that describes, under the conditions given, an upper bound on the error of estimation for the three methods in this section.

Rule	**Condition for all $x \in [a, b]$**	**Maximum Error**
Midpoint	$\left|f''(x)\right| \le M$	$\frac{M(b-a)^3}{24n^2}$
Trapezoidal	$\left|f''(x)\right| \le M$	$\frac{M(b-a)^3}{12n^2}$
Simpson's	$\left|f^{(4)}(x)\right| \le M$	$\frac{M(b-a)^5}{180n^4}$

5)   Find the maximum error in estimating $\int_{-1}^{3} x^3 \, dx$ with $n = 8$ using the Trapezoidal Rule.

The maximum is $\frac{M(3-(-1))^3}{12(8)^2} = \frac{M}{12}$.

To find a value for M we note $f''(x) = 6x$. For $-1 \le x \le 3$, $\left|f''(x)\right| = \left|6x\right| \le 6 \cdot 3 = 18$. We choose $M = 18$. Thus the maximum error of estimation is $\frac{M}{12} = \frac{18}{12} = 1.5$.

6)   Find the maximum error in estimating $\int_{1}^{3} \sin 2x \, dx$ with $n = 4$ using Simpson's Rule.

$\frac{M(3-1)^3}{180(4)^4} = \frac{M}{1440}$.

For $f(x) = \sin 2x$, $f^{(4)}(x) = 16 \sin 2x$. For $1 \le x \le 3$, $\left|\sin 2x\right| \le 1$, so $\left|f^{(4)}(x)\right| \le 2^4 = 16$. We choose $M = 16$.

Thus the maximum error is $\frac{M}{1440} = \frac{16}{1440} = \frac{1}{90}$.

Section 7.9

Improper Integrals

Concepts to Master

A. Definition and evaluation of improper integrals; Type1; Type 2; Convergent and Divergent

B. Comparison Test for integrals

Summary and Focus Questions

A. Improper integrals take several forms. They may have an infinite limit of integration, as in $\int_1^\infty \frac{1}{x^2}\,dx$. This is a Type 1 integral. They may have a "trouble point" where the function is not defined or grows large, as in $\int_0^1 \frac{1}{x^2}\,dx$ or $\int_{-1}^1 \frac{1}{x^2}\,dx$ $\left(\frac{1}{x^2}\text{ is not defined at zero}\right)$. This is a Type 2 integral.

The two types of improper integrals (three cases each) defined below are each expressed as a limit of ordinary integrals.

Definition	**Necessary Condition**
Type 1	
a) $\int_a^\infty f(x)\,dx = \lim\limits_{t\to\infty}\int_a^t f(x)\,dx$	$\int_a^t f(x)\,dx$ exists for all $t \geq a$
b) $\int_{-\infty}^b f(x)\,dx = \lim\limits_{t\to-\infty}\int_t^b f(x)\,dx$	$\int_t^b f(x)\,dx$ exists for all $t \leq b$
c) $\int_{-\infty}^\infty f(x)\,dx = \int_{-\infty}^a f(x)\,dx + \int_a^\infty f(x)\,dx$	Both integrals on the right exist
Type 2	
a) $\int_a^b f(x)\,dx = \lim\limits_{t\to a^+}\int_t^b f(x)\,dx$	$\int_t^b f(x)\,dx$ exists for $t \in (a,\,b]$
b) $\int_a^b f(x)\,dx = \lim\limits_{t\to b^-}\int_a^t f(x)\,dx$	$\int_a^t f(x)\,dx$ exists for all $t \in [a,\,b)$
c) $\int_a^b f(x)\,dx = \int_a^c f(x)\,dx + \int_c^b f(x)\,dx$	Both improper integrals on the right exist

If an improper integral exists, it is said to <u>converge</u>; otherwise it <u>diverges</u>. Improper integrals are evaluated as limits according to their definitions.

1) Give a definition of each:

a) $\int_0^\infty \sin 2x \, dx$

$\lim_{t\to\infty} \int_0^t \sin 2x \, dx$

b) $\int_0^1 \frac{1}{\sqrt{x}} \, dx$

$\lim_{t\to 0^+} \int_t^1 \frac{1}{\sqrt{x}} \, dx$

c) $\int_{-\infty}^\infty e^{-x^2} \, dx$

$\int_{-\infty}^0 e^{-x^2} \, dx + \int_0^\infty e^{-x^2} \, dx$ (The choice of zero was arbitrary; any real number would do.)

d) $\int_{-2}^2 \frac{1}{(x-1)^2} \, dx$

$\int_{-2}^1 \frac{1}{(x-1)^2} \, dx + \int_1^2 \frac{1}{(x-1)^2} \, dx$

e) $\int_1^\infty \frac{1}{1-x} \, dx$

This one is both types and should be rewritten as
$\int_1^2 \frac{1}{1-x} \, dx + \int_2^\infty \frac{1}{1-x} \, dx$
$= \lim_{t\to 1^+} \int_t^2 \frac{1}{1-x} \, dx + \lim_{t\to\infty} \int_2^t \frac{1}{1-x} \, dx.$

2) Evaluate $\int_0^{\pi/2} \sec^2 x \, dx$

$\int_0^{\pi/2} \sec^2 x \, dx = \lim_{t\to\pi/2^-} \int_0^t \sec^2 x \, dx$
$= \lim_{t\to\pi/2^-} \tan x \Big|_0^t = \lim_{t\to\pi/2^-} \tan t$
which does not exist (the limit is ∞). This integral diverges.

3) Evaluate $\int_1^\infty x e^{-x^2} \, dx$

$\int_1^\infty x e^{-x^2} \, dx = \lim_{t\to\infty} \int_1^t x e^{-x^2} \, dx$
$(u = -x^2, \, du = -2x \, dx)$
$= \lim_{t\to\infty} \left(-\tfrac{1}{2} e^{-x^2} \Big|_1^t \right)$
$= \lim_{t\to\infty} \left(-\tfrac{1}{2} (e^{-t^2} - e^1) \right)$
$= -\tfrac{1}{2}(0 - e) = \tfrac{e}{2}.$

4) a) Find the area of the shaded
region.

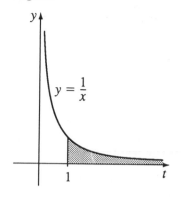

b) Evaluate $\int_1^\infty \frac{\pi}{x^2}\, dx$

c) (for fun) The region in part a) has
an infinite area. How can it be
painted with a finite amount of
paint?

$\int_1^\infty \frac{1}{x}\, dx = \lim_{t\to\infty} \int_1^t \frac{1}{x}\, dx$

$= \lim_{t\to\infty} \ln t \Big|_1^t = \lim_{t\to\infty} \ln t = \infty.$

The improper integral does not exist; the
area is infinite.

$\int_1^\infty \frac{\pi}{x^2}\, dx = \lim_{t\to\infty} \int_1^t \int_1^t \frac{\pi}{x^2}\, dx$

$= \lim_{t\to\infty} \frac{-\pi}{x} \Big|_1^t = \lim_{t\to\infty} \frac{-\pi}{t} + \frac{\pi}{1} = \pi.$

Step 1. Take the region, $0 \le y \le \frac{1}{x}$,
$1 \le x \le \infty$, and revolve it about
the x-axis.
Step 2. The volume of the resulting solid
of revolution is given by $\int_1^\infty \frac{\pi}{x^2}\, dx$,
which by part b) is π.
Step 3. Fill the solid of revolution with π
gallons of paint.
Step 4. Dip the region in part a) in the
solid of revolution. The region is
then painted with at most π gallons
of paint!

B. Sometimes improper integrals can be shown to converge without actually
determining the exact value it converges to. (It is useful information to know
an improper integral exists when it will be approximated, for instance.) One
such method to determine existance is the <u>Comparison Test for Improper
Integrals</u>.

Suppose $f(x)$ and $g(x)$ are continuous and $f(x) \geq g(x) \geq 0$ for $x \geq a$.
If $\int_a^\infty f(x)dx$ converges then $\int_a^\infty g(x)dx$ converges.
Similar statements can be made for the other forms of improper integrals.

5) True or False:
    If $f(x) \geq g(x) \geq 0$ for $x \geq a$ and f
    and g are continuous, then if $\int_a^b g(x)dx$
    diverges, then $\int_a^\infty f(x)dx$ diverges.

True. This statement is just the contrapositive of the Comparison Test for Improper Integrals.

6) Show that $\int_1^\infty \frac{|\sin x|}{x^2}\, dx$ converges.

For $1 \leq x \leq \infty$, $0 \leq |\sin x| \leq 1$.
Thus $0 \leq \frac{|\sin x|}{x^2} \leq \frac{1}{x^2}$.
$\int_1^\infty \frac{1}{x^2}\, dx$ converges (to 1).
Thus $\int_1^\infty \frac{|\sin x|}{x^2}\, dx$ converges, (but we don't know to what value, except that it will be between 0 and 1.)

8

Further Applications of Integration

"THAT WRAPS IT UP — THE MASS OF THE UNIVERSE."

Cartoons courtesy of Sidney Harris. Used by permission.

Differential Equations

Concepts to Master

A. Order, degree, general and particular solution to a differential equation; Separable equations
B. Directional fields

Summary and Focus Questions

A. A <u>differential equation</u> is an equation involving x, y, y', y'', $\ldots$, $y^{(n)}$. The n in the highest $y^{(n)}$ is the <u>order</u> of the equation and the <u>degree</u> is the exponent that $y^{(n)}$ has. For example, $2(y''')^4 + 5xy' + 7x = 0$ has degree 4 and order 3.

A <u>particular solution</u> to a differential equation is a function $y = f(x)$ that satisfies the equation. A <u>general solution</u> is an expression with arbitrary constants to represent all particular solutions. <u>Initial boundary conditions</u> are values used to determine a particular solution from the general. For example, the equation $y' = 10x$ has general solution $y = 5x^2 + C$. If we specify an initial condition of $y(1) = 12$, then $12 = 5 + C$ implies $C = 7$ so the particular solution for the condition $y(1) = 12$ is $y = 5x^2 + 7$.

A <u>separable differential equation</u> is an equation that can be written as $\frac{dy}{dx} = \frac{f(x)}{g(y)}$. It is solved by first rewriting as $g(y)dy = f(x)dx$, then integrating to get $\int g(y)dy = \int f(x)dx + C$, and finally solving this equation for y.

1) The equation
$$x^2(y'')^3 + 4xy' - 2y + x = 0 \text{ has}$$
degree _____ and order _____.

degree 3, order 2

2) Is $y' = xy + x$ separable?

Yes. $\frac{dy}{dx} = x(y+1) = \frac{x}{(y+1)^{-1}}$.

3) Solve the equation $y' = \frac{\sin x}{\cos y}$.

The equation is separable.
$$\frac{dy}{dx} = \frac{\sin x}{\cos y}$$
$$\cos y\, dy = \sin x\, dx$$
$$\int \cos y\, dy = \int \sin x\, dx$$
$$\sin y = -\cos x + C$$
$$y = \sin^{-1}(-\cos x + C).$$

4) Find the particular solution for the initial condition $y(0) = 2$ for the equation $e^{x^2} yy' + x = 0$.

First separate: $e^{x^2} y \frac{dy}{dx} = -x$
$$y\frac{dy}{dx} = -xe^{-x^2}$$
Integrate: $\int y\, dy = \int -xe^{-x^2}\, dx$

$$\frac{y^2}{2} = \frac{1}{2}e^{-x^2} + C$$
$$y^2 = e^{-x^2} + C$$
$$y = (e^{-x^2} + C)^{1/2} = \sqrt{e^{-x^2} + C}$$
Find the particular solution:
At $x = 0$, $y = 2$ so $\sqrt{e^{-0} + C} = 2$.
Thus $1 + C = 4$, $C = 3$.
The particular solution is $y = \sqrt{e^{-x^2} + 3}$.

B. The general solution to a differential equation $y' = F(x, y)$ produces a family of functions whose graphs are related. One means to sketch these graphs is to draw a directional field of short line segments with slopes $F(x, y)$ at (x, y). A table of calculated values helps:
For example $y' = x + y - 1$ yields this table

x	-1	-1	-1	0	0	0	1	1	1	2	2	2
y	0	1	2	0	1	2	0	1	2	0	1	2
y'	-2	-1	0	-1	0	1	0	1	2	1	2	3

and this directional field.

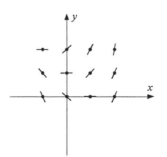

which suggests this family of graphs for the general solution (In chapter 15 we will see the solution is $y = -x + Ce^x$.)

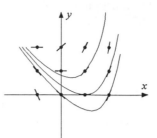

5) Find the directional field and sketch the general solution to $y' = xy + y$.

x	y	y'
-2	-1	1
-2	0	0
-2	1	-1
-1	-1	0
-1	0	0
-1	1	0
0	-1	-1
0	0	0
0	1	1
1	-1	-2
1	0	0
1	1	2

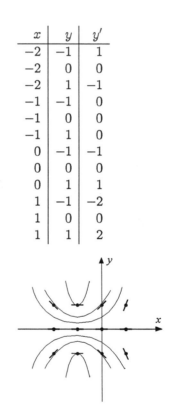

Note: The equation is separable and has general solution $y = Ce^{(x^2/2)+x}$.

Arc Length

Concepts to Master

A. Length of a curve
B. Arc length function and its derivative

Summary and Focus Questions

A. If f' is continuous on $[a, b]$, the <u>length of the curve</u> $y = f(x)$, $a \le x \le b$ (arc length) is

$$L = \int_a^b \sqrt{1 + (f'(x))^2}\ dx.$$

For curves described as $x = g(y)$ for $c \le y \le d$, the arc length is
$\int_c^d \sqrt{1 + g'(y))^2}\ dy$.

1) Find a definite integral for the length of $y = \frac{1}{x}$, $1 \le x \le 3$.

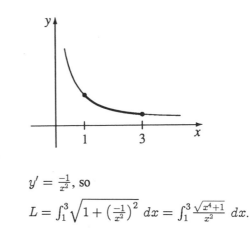

$y' = \frac{-1}{x^2}$, so

$$L = \int_1^3 \sqrt{1 + \left(\frac{-1}{x^2}\right)^2}\ dx = \int_1^3 \frac{\sqrt{x^4 + 1}}{x^2}\ dx.$$

2) What is the length of the arc given by $x = 1 + y^{3/2}$, $0 \le y \le 4$?

$x' = \frac{3}{2}y^{1/2}$

$L = \int_0^4 \sqrt{1 + \left(\frac{3}{2}y^{1/2}\right)^2}\ dy$

$\qquad = \int_0^4 \sqrt{1 + \frac{9}{4}y}\ dy$

$\qquad \left(u = 1 + \frac{9}{4}y,\ du = \frac{9}{4}\ dy\right)$

$\qquad = \frac{4}{9}\int \sqrt{u}\ du = \frac{4}{9}\frac{u^{3/2}}{3/2} = \frac{8}{27}u^{3/2}$

$\qquad = \frac{8}{27}\left(1 + \frac{9}{4}y\right)^{3/2}\Big|_0^4$

$\qquad = \frac{8}{27}(10\sqrt{10} - 1).$

3) Write an expression involving definite integrals for the length of the curve $y = \sqrt[3]{x}$ from $x = -1$ to $x = 8$.

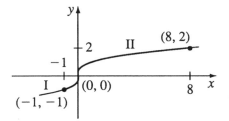

Since a vertical tangent exists at $x = 0$, the arc length expression $\int_{-1}^8 \ldots\ dx$ would result in an improper integral. We could divide the curve at $(0, 0)$ into arc I and arc II. The curve length is the sum of the lengths of these two arcs. Instead, we choose to treat x as a function of y. From $y = \sqrt[3]{x}$, $x = y^3$ and $x' = 3y^2$. The arc length is

$\int_{-1}^2 \sqrt{1 + (3y^2)^2}\ dy = \int_{-1}^2 \sqrt{1 + 9y^4}\ dy.$

B. For a smooth function $y = f(x)$, $a \le x \le b$, let $s(x)$ be the distance along the curve from $(a,\ f(a))$ to $(x,\ f(x))$. The function $s(x)$ may be written as

$$s(x) = \int_a^x \sqrt{1 + [f'(t)]^2}\ dt.$$

The differential $ds = \sqrt{1 + \left(\frac{dy}{dx}\right)^2}\, dx$ may be used to express arc length
$$L = \int ds.$$

This form is handy because, from $ds = \sqrt{1 + \left(\frac{dy}{dx}\right)^2}$ we can write
$$(ds)^2 = (dx)^2 + (dy)^2$$

This last form can be used to remember <u>both</u>
$$ds = \sqrt{1 + \left(\frac{dy}{dx}\right)^2}\, dx \text{ and } ds = \sqrt{1 + \left(\frac{dx}{dy}\right)^2}\, dy.$$

4) Find the arc length function for
$y^2 = 4x^3$, $0 \leq x < 1$.

$y^2 = 4x^3$, $y = 2x^{3/2}$
$\frac{dy}{dx} = 3x^{1/2}$
$s(x) = \int_0^x \sqrt{1 + (3t^{1/2})^2}\, dt$
$\qquad = \int_0^x \sqrt{1 + 9t}\, dt$
$\qquad\quad (u = 1 + 9t,\ du = 9\, dt)$
$\qquad = \frac{1}{9}\frac{(1+9t)^{3/2}}{3/2} = \frac{2}{27}(1 + 9t)^{3/2}\Big|_0^x$
$\qquad = \frac{2}{27}(1 + 9x)^{3/2} - \frac{2}{27}.$

Section 8.3

Area of a Surface of Revolution

Concepts to Master

Computation of surface area of a solid of revolution

Summary and Focus Questions

If $f(x) \geq 0$ and $f'(x)$ is continuous on $[a, b]$, the <u>surface area</u> of the solid of revolution obtained by rotating $f(x)$ about the x-axis is

$$S = \lim_{\|P\| \to 0} \sum_{i=1}^{n} 2\pi f(x_i^*) \sqrt{1 + (f'(x_i^*))^2} \, \Delta x_i$$

$$= \int_a^b 2\pi f(x) \sqrt{1 + (f'(x))^2} \, dx.$$

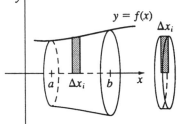

For curves described by $x = g(y)$, $c \leq y \leq d$,

$$S = \int_c^d 2\pi y \sqrt{1 + \left(\frac{dx}{dy}\right)^2} \, dy.$$

Recalling that $(ds)^2 = (dx)^2 + (dy)^2$, more compact notation may be used which represents both the above cases of rotation about the x-axis.

$$S = \int 2\pi y \, ds, \text{ where } ds = \sqrt{1 + \left(\frac{dy}{dx}\right)^2} \, dx \text{ or } ds = \sqrt{1 + \left(\frac{dx}{dy}\right)^2} \, dy$$

For rotation about the y-axis, $S = \int 2\pi x \, ds$.

1) Find the surface area of the solid obtained by revolving $y = \sin 2x$, $0 \leq x \leq \frac{\pi}{2}$, about the x-axis.

$$S = \int_0^{\pi/2} 2\pi \sin 2x \sqrt{1 + (2\cos 2x)^2} \, dx$$

$$\left(\begin{array}{l} u = 2\cos 2x, \, du = -4\sin 2x \\ \text{At } x = 0, \, u = 2; \, x = \pi/2, \, u = -2 \end{array} \right)$$

$$= \frac{-\pi}{2} \int_2^{-2} \sqrt{1 + u^2} \, du = \frac{\pi}{2} \int_{-2}^{2} \sqrt{1 + u^2} \, du$$

(Table of Integrals, #21)

$$= \frac{u}{2}\sqrt{1+u^2} + \frac{1}{2}\ln\left|u + \sqrt{1+u^2}\right| \Big|_{-2}^{2}$$

$$= \frac{2}{2}\sqrt{5} + \frac{1}{2}\ln\left|2 + \sqrt{5}\right|$$
$$\quad -\left(\frac{-2}{2}\sqrt{5} + \frac{1}{2}\ln\left|-2 + \sqrt{5}\right|\right)$$

$$= 2\sqrt{5} + \frac{1}{2}\ln\left(2 + \sqrt{5}\right)$$
$$\quad - \frac{1}{2}\ln\left(-2 + \sqrt{5}\right).$$

2) Set up two equivalent definite integrals for the surface area of the solid obtained by revolving $y = x^2$, $0 \le x \le 2$ about the y-axis.

In both solutions $s = \int 2\pi x \, ds$.

I. $ds = \sqrt{1 + \left(\frac{dy}{dx}\right)^2} \, dx$.

$y = x^2$ so $\frac{dy}{dx} = 2x$.

$$S = \int_0^2 2\pi x \sqrt{1 + (2x)^2} \, dx$$

$$\quad = \int_0^2 2\pi x \sqrt{1 + 4x^2} \, dx$$

II. $ds = \sqrt{1 + \left(\frac{dy}{dx}\right)^2} \, dy$.

Since $y = x^2$, $0 \le x \le 2$,
$x = \sqrt{y}$, $0 \le y \le 4$ and $\frac{dx}{dy} = \frac{1}{2\sqrt{y}}$.

$$S = \int_0^4 2\pi \sqrt{y} \sqrt{1 + \left(\frac{1}{2\sqrt{y}}\right)^2} \, dy$$

$$\quad = \int_0^4 2\pi \sqrt{y} \sqrt{\frac{1+4y}{4y}} \, dy$$

$$\quad = \int_0^4 \pi \sqrt{1 + 4y} \, dy.$$

The integral in part II is equivalent to the integral in part I since $y = x^2$, $dy = 2x \, dx$, and for $0 \le x \le 2$, we have $0 \le y \le 4$.

Moments and Centers of Mass

Concepts to Master

A. Moments about x-axis and y-axis; Centroid (center of mass)

B. Theorem of Pappus

Summary and Focus Questions

A. A lamina is a thin flat sheet of material determined by the region R, in the figure whose density is uniformly ρ, a constant.

Let $A = \int_a^b f(x)\,dx$ be the area of R. The <u>centroid</u> (or <u>center of mass</u>) of R is the point $(\overline{x}, \overline{y})$ where

$\overline{x} = \frac{1}{A}\int_a^b xf(x)\,dx$ and
$\overline{y} = \frac{1}{A}\int_a^b \frac{1}{2}[f(x)]^2\,dx$.

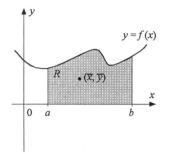

If the lamina was suspended in the air by a string attached at the centroid $(\overline{x}, \overline{y})$, then it would balance.

The <u>moment of R about the x-axis</u> is $M_x = \rho\int_a^b \frac{1}{2}[f(x)]^2\,dx$ and the <u>moment about the y-axis</u> is $M_y = \rho\int_a^b xf(x)dx$. These are measures of tendency to rotate about the x-axis and y-axis, respecitively. Note that $\overline{x} = \frac{M_y}{\rho A}$ and $\overline{y} = \frac{M_x}{\rho A}$.

1) Find M_x, M_y, and the center of mass of the region bounded by $y = 4 - x$, $x = 0$, $y = 0$, with density ρ.

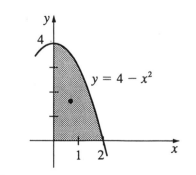

The area of the region is

$A = \int_0^2 (4 - x^2)dx = 4x - \frac{x^3}{3} \Big|_0^2 = \frac{16}{3}$.

For M_y and $\overline{x}$:

$\int_a^b x f(x)dx = \int_0^2 x(4 - x^2)dx$

$\qquad = \int_0^2 (4x - x^3)dx$

$\qquad = \left(2x^2 - \frac{x^4}{4}\right)\Big|_0^2 = 4$.

Therefore, $M_y = 4\rho$ and

$\overline{x} = \frac{1}{16/3\rho}(4\rho) = \frac{3}{4}$.

For M_x and $\overline{y}$:

$\int_a^b \frac{1}{2}[f(x)]^2\ dx = \frac{1}{2}\int_0^2 (4 - x^2)^2\ dx$

$\qquad = \frac{1}{2}\int_0^2 (16 - 8x^2 + x^4)dx$

$\qquad = \frac{1}{2}\left(16x - \frac{8}{3}x^3 + \frac{x^5}{5}\right)\Big|_0^2 = \frac{128}{15}$.

Therefore $M_x = \frac{128}{15}\rho$ and

$\overline{y} = \frac{1}{16/3\rho}\left(\frac{128}{15}\rho\right) = \frac{8}{5}$.

B. The <u>Theorem of Pappus</u> is: If a region R is revolved about a line L that does not intersect R, the resulting solid has volume $V = 2\pi d A$ where A is the area of R and d is the perpendicular distance between the centroid of R and the line L. (Notice that $2\pi d$ is the distane traveled by the centroid.)

Be careful that you do not interpret dA as a differential; in this case dA means the product of the distance and the area.

2) Find the volume of the solid of revolution obtained by rotating the region in question 1 about:

a) the x-axis.

$\overline{y} = \frac{8}{5}$ is distance from the center of mass to the axis of revolution (the x-axis). The area is $A = \frac{16}{3}$.

Thus $V = 2\pi\left(\frac{8}{5}\right)\left(\frac{16}{3}\right) = \frac{256\pi}{9}$.

b) the y-axis.

$\overline{x} = \frac{3}{4}$, so $V = 2\pi\left(\frac{3}{4}\right)\left(\frac{16}{3}\right) = 8\pi$

Hydrostatic Pressure and Force

Concepts to Master

Calculation of force on a vertical plate due to hydrostatic pressure; Calculation of pressure at a given depth

Summary and Focus Questions

The pressure on a plate suspended horizontally in a liquid with density ρ at a depth d is

$$P = \rho g d,$$

where g is the gravitational constant. P is in units of newtons per meter^2 (= pascals) or in pounds per ft^2.

Suppose a plate is suspended vertically in a liquid with mass density ρ at depth H and the surface of the place can be described as bounded by $x = f(y)$, $x = k(y)$, $y = c$, $y = d$ with $f(y) \geq k(y)$ for all $y \in [c, d]$. Then the force due to liquid pressure on a section of the plate is approximately

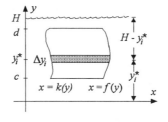

(density)(gravitational constant)(depth)(area of section)
$$= \rho g(H - y_i^*)[f(y_i^*) - k(y_i^*)]\Delta y_i$$

Thus the <u>total hydrostatic force on the surface</u> is $\int_c^d \rho g(H - y)(f(y) - k(y))dy$.

1) A circular plate of radius 3 cm is suspended horizontally at a depth of 12 m in an oil having density 1500 kg/m^3. What is the pressure on the plate?

Since the plate is horizontal, the pressure is unform at all points.
$$P = \rho g d = (1500 \text{ kg/m}^3)(9.8 \text{ m/s}^2)(12 \text{ m})$$
$$= 176,400 \text{ Pa (pascals)} = 176.4 \text{ kPa}.$$

2) A flat isoceles triangle is suspended in water as in the figure. Water density is 1000 kg/m³. Set up a definite integral for the total hydrostatic force.

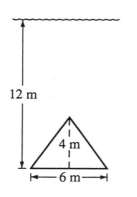

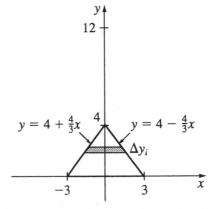

Setting up coordinates a shown the sides of the triangle are
$y = 4 - \frac{4}{3}x$, so $x = \frac{-3}{4}y + 3$, and
$y = 4 + \frac{4}{3}x$, so $x = \frac{3}{4}y - 3$.
The force on the triangle is

$$\int_0^4 1000(9.8)(12 - y)\left[\left(-\tfrac{3}{4}y + 3\right)\right.$$

$$\left. - \left(\tfrac{3}{4}y - 3\right)\right] dy$$
$$= \int_0^4 9800(12 - y)\left(6 - \tfrac{3}{2}y\right) dy.$$

Applications to Economics and Biology

Concepts to Master

Using a definite integral to determine the total value of a quantity

Summary and Focus Questions

Applications of definite integrals in this section and elsewhere all involve calculating the total amount of a quantity (that can be represented by a continuous $y = f(x)$) within a certain range ($a \le x \le b$). The idea of summing the values (Riemann sums) is extended to the continuous case $\left(\lim\limits_{\|P\| \to 0} \right)$ to give the definite integral ($\int_a^b f(x)dx$). Here are two examples from economics:

1. A demand function $p(x)$ is the price necessary to sell x items. Let X be the amount actually available and P be the price (constant) for that amount, $P = p(X)$. The <u>consumer's surplus</u> is
$$\int_0^X [p(x) - P]dx.$$
This is the total amount of money that could be saved by consumers if they buy when the price is P dollars.

2. If $f(t)$ is the amount of money that is to be received at time $t \in [a, b]$ and interest is compounded continuously at $r\%$, the <u>present value</u> of $f(t)$ is
$$\int_a^b e^{-rt} f(t)dt.$$
This is the total amount of money that must be invested now so that the amounts collected at times $t(a \le t \le b)$ will be $f(t)$ dollars.

1) Find a definite integral for the consumer surplus determined by 1000 units available and a demand function of $p(x) = 85 - 0.03x$.

At $X = 1000$,
$P = 85 - 0.03(1000) = 75$.
The consumer surplus is
$\int_0^{1000} [85 - 0.03x - 75]dx$
$= \int_0^{1000} (10 - 0.03x)dx$

2) A family sets up a trust fund for a 10-year-old daughter so that between the ages of 10 and 21 she will receive $1000 the first year and double the amount each following year. Find a definite integral for how much needs to be invested now (what is the present value) if interest is compounded continuously at 7%?

$f(t) = 1000\, 2^t$, $t \in [0, 11]$.
The present value is $\int_0^4 e^{-0.07t}(1000\, 2^t)dt$

9

Parametric Equations and Polar Coordinates

Cartoons courtesy of Sidney Harris. Used by permission.

Curves Defined by Parametric Equations

Concepts to Master

Parametric equations; Graphs of curves defined by parametric equations; Elimination of the parameter

---

Summary and Focus Questions

A set of <u>parametric equations</u> has the form
$$x = f(t), \; y = g(t)$$
where f and g are functions of a third variable t called a <u>parameter</u>. To each t there corresponds an x value and a y value, hence a point in the xy plane. The collection of all such points is a <u>curve</u>. The same curve may be described by several different pairs of equations.

Some pairs of parametric equations can be reduced algebraically to a single equation not involving t; this is called <u>eliminating the parameter</u>. The method to do so depends greatly on the nature of the functions involved. Often one equation can be solved for t and the result substituted for t in the other.

<u>Example</u>: Eliminate the parameter in $x = 2 + 3t$, $y = t^2 - 2t + 1$.
Solve $x = 2 + 3t$ for t to get $t = \frac{1}{3}(x - 2)$. Substitute this in the other equation:
$$y = \left[\tfrac{1}{3}(x - 2)\right]^2 - 2\left[\tfrac{1}{3}(x - 2)\right] + 1 \; or \; y = \tfrac{1}{9}x^2 - \tfrac{10}{9}x + \tfrac{25}{9}.$$
If trigonometric functions are involved, look for a trigonometric identity to help eliminate t.

<u>Example</u>: Eliminate the parameter in $x = \sin t$, $y = \cot^2 t$.
$$y = \cot^2 t = \frac{\cos^2 t}{\sin^2 t} = \frac{1 - \sin^2 t}{\sin^2 t}. \; \text{Thus } y = \frac{1 - x^2}{x^2}.$$
We note from $x = \sin t$, $-1 \leq x \leq 1$ and $x \neq 0$ (since $\cot t$ is undefined when $\sin t = 0$.)

1) Sketch a graph of the curve given by
   $x = \sqrt{t}$, $y = t + 2$.

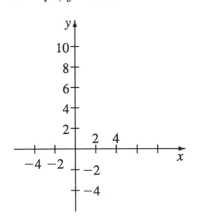

Compute some values and plot points:

t	0	1	4	9
x	0	1	2	3
y	2	3	6	11

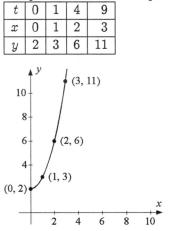

We see this is half a parabola since
$x = \sqrt{t}$ implies $x^2 = t$, thus $y = x^2 + 2$,
$x \geq 0$.

2) Eliminate the parameter in each

   a) $x = e^t$, $y = t^2$

Solve for t in terms of y:
$y = t^2$, $\sqrt{y} = t$.
Thus $x = e^{\sqrt{y}}$, $x > 0$.
A solution may also be obtained by solving
for t in terms of x:
$x = e^t$, $t = \ln x$, $y = (\ln x)^2$, $x > 0$.

   b) $x = 1 + \cos t$, $y = \sin^2 t$

$x = 1 + \cos$, $x - 1 = \cos t$
$\cos^2 t = (x - 1)^2$.
Since $\sin^2 t = y$, and $\cos^2 t + \sin^2 t = 1$
we have $(x - 1)^2 + y = 1$.
Thus $y = 1 - (x - 1)^2$.
Since $x = 1 + \cos t$, $0 \leq x \leq 2$.

c) $x = 2\sec t$, $y = 3\tan t$

$\sec t = \frac{x}{2}$ and $\tan t = \frac{y}{3}$.

Hence the identity $\tan^2 t + 1 = \sec^2 t$
becomes $\left(\frac{y}{3}\right)^2 + 1 = \left(\frac{x}{2}\right)^2$ or
$\frac{x^2}{4} - \frac{y^2}{9} = 1$.

3) Describe the graph of $x = a + bt$,
$y = c + dt$ where a, b, c, d are
constants.

If both b and d are zero, this is the single
point (a, c). If $b = 0$ and $d \neq 0$ this is a
vertical line through (a, c). If $b \neq 0$, then
$t = \frac{x-a}{b}$ and $y = c + d\frac{(x-a)}{b}$. Thus
$y = \frac{d}{b}x + c - \frac{da}{b}$, so the graph is a line
through (a, c) with slope $\frac{d}{b}$.

Tangents and Areas

Concepts to Master

A. First and second derivative of a function defined by a pair of parametric equations

B. Area under a curve defined by a pair of parametric equations

Summary and Focus Questions

A. The slope of the line tangent to a curve described by $x = f(t)$, $y = g(t)$ is the first derivative of y with respect to x and is given by

$$\frac{dy}{dx} = \frac{\frac{dy}{dt}}{\frac{dx}{dt}} \text{ if } \frac{dx}{dt} \neq 0.$$

In the case $\frac{dx}{dt} = 0$ but $\frac{dy}{dt} \neq 0$, the curve will have a vertical tangent line. The second derivative of y with respect to x is:

$$\frac{d^2 y}{dx^2} = \frac{\frac{d}{dt}\left(\frac{dy}{dx}\right)}{\frac{dx}{dt}}.$$

1) For $x = t^3$, $y = t^2 - 2t$

   a) Find $\frac{dy}{dx}$ and $\frac{d^2 y}{dx^2}$.

   $\frac{dx}{dt} = 3t^2$ and $\frac{dy}{dt} = 2t - 2$, so $\frac{dy}{dx} = \frac{2t-2}{3t^2}$.

   $\frac{d}{dt}\left(\frac{dy}{dx}\right) = \frac{3t^2(2) - (2t-2)(6t)}{(3t^2)^2} = \frac{4-2t}{3t^3}$.

   Thus $\frac{d^2 y}{dx^2} = \frac{\frac{4-2t}{3t^3}}{3t^2} = \frac{4-2t}{9t^5}$.

   b) Find the horizontal and vertical tangents for the curve.

   $\frac{dy}{dx} = 0$ at $t = 1$ $(x = 1, y = -1)$.
   A horizontal tangent is at $(1, -1)$.
   $\frac{dy}{dx}$ does not exist at $t = 0$ $(x = 0, y = 0)$.
   A vertical tangent is at $(0, 0)$.

c) Discuss the concavity of the
curve.

$\frac{d^2y}{dx^2} > 0$ for $0 < t < 2$ $(0 < x < \sqrt[3]{2})$ and
negative for $t < 0$ $(x < 0)$ and $t > 1 - 2$
$(x > \sqrt[3]{2})$. The curve is concave upward
for $x \in (0, \sqrt[3]{2})$ and concave downward
for $x \in (-\infty, 0)$ and $x \in (\sqrt[3]{2}, \infty)$.

d) Sketch the curve.

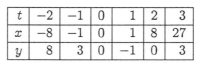

t	-2	-1	0	1	2	3
x	-8	-1	0	1	8	27
y	8	3	0	-1	0	3

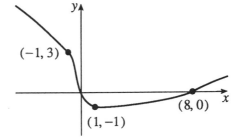

2) Find the equation of the tangent line
to the curve $x = 4t^2 + 2t + 1$,
$y = 7t + 2t^2$ at the point
corresponding to $t = -1$.

At $t = -1$, $x = 3$ and $y = -5$.
$\frac{dx}{dt} = 8t + 2 = -6$ at $t = -1$.

$\frac{dy}{dt} = 7 + 4t = 3$ at $t = -1$.
Thus $\frac{dy}{dx} = \frac{3}{-6} = -\frac{1}{2}$. The tangent line is
$y + 5 = -\frac{1}{2}(x - 3)$.

B. Suppose the parametric equations $x = f(t)$, $y = g(t)$, $t \in [t_1, t_2]$ define an
integrable function $y = F(x) \geq 0$ over the interval $[a, b]$, where $a = f(t_1)$,
$b = f(t_2)$. The area under $y = F(x)$ is $\int_a^b y \, dx = \int_{t_1}^{t_2} g(t) f'(t) dt$.

Using the ds notation this is $S = \int 2\pi y \, ds$, the same formula as in Section 8.3.

2) Find a definite integral for the area of the surface of revolution about the x-axis for each curve:

a) $x = 2t + 1$, $y = t^3$, $t \in [0, 2]$

$$S = \int_0^2 2\pi(t^3)\sqrt{(2)^2 + (3t^2)^2} \, dt$$
$$= \int_0^2 2\pi t^3 \sqrt{4 + 9t^4} \, dt.$$

b) a "football" obtained from rotating the ellipse $\frac{x^2}{a^2} + \frac{y^2}{b^2} = 1$, $y \geq 0$ about the x-axis.

The top half of the ellipse is $x = a \cos t$, $y = b \sin t$, $t \in [0, \pi]$.
$$ds = \sqrt{(dx)^2 + (dy)^2}$$
$$= \sqrt{(-a \sin t)^2 + (b \cos t)^2}$$
$$= \sqrt{(a^2 \sin^2 t + b^2 \cos^2 t}.$$
Thus the area
$$S = \int_0^\pi 2\pi(b \sin t)\sqrt{a^2 \sin^2 t + b^2 \cos^2 t} \, dt.$$

Polar Coordinates

Concepts to Master

A. Plotting points in polar coordinates; Conversion to and from rectangular coordinates

B. Graphs of polar equations; Tangents to polar curves

Summary and Focus Questions

A. To construct a <u>polar coordinate system</u> start with a point called the <u>pole</u> and a ray from the pole called the <u>polar axis</u>. Then a point P has polar coordinates (r, θ) if

$|r|$ = the distance from the pole to P, and
θ = the measure of a directed angle with initial side the polar axis and terminal side the line through the pole and P.

To plot a point P with polar coordinates $P(r, \theta)$:
if $r = 0$, P is the pole (for any value of θ).

if $r > 0$, rotate the polar axis by the angle θ (counterclockwise for $\theta > 0$, clockwise for $\theta < 0$) and locate P on this ray at a distance r units from the pole.

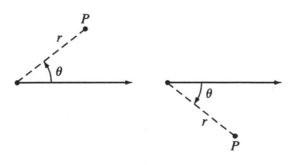

237

if $r < 0$, rotate the polar axis by the angle θ and then reflect about the pole. P is located on this reflected ray at a distance $-r$ units from the pole.

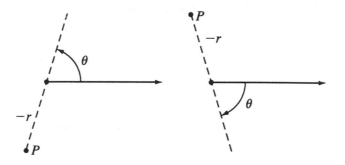

If a rectangular coordinate system is placed upon the polar coordinate system as in the figure, then to change from polar to rectangular:

$x = r\cos\theta$
$y = r\sin\theta$

to change from rectangular to polar:

θ is a solution to $\tan\theta = \frac{y}{x}$ (for $x \neq 0$).
r is a solution to $r^2 = x^2 + y^2$ where
$r > 0$ if the terminal side of the angle
θ is in the same quadrant as P;
if not, then $r \leq 0$.

1) Plot these points in polar coordinates.

A: $\left(4, \frac{\pi}{3}\right)$ B: $\left(-3, \frac{\pi}{4}\right)$
C: $(0, 3\pi)$ D: $\left(2, -\frac{\pi}{4}\right)$

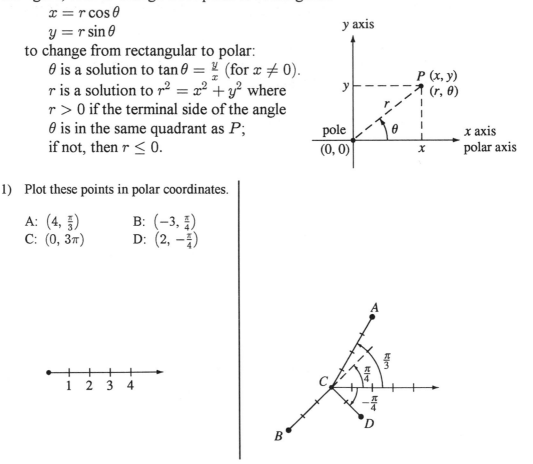

2) Find two other polar coordinates for point with polar coordinates $\left(8, \frac{\pi}{3}\right)$.

There are infinitely many answers, including $\left(8, \frac{7\pi}{3}\right)$, $\left(-8, \frac{4\pi}{3}\right)$, $\left(8, \frac{-5\pi}{3}\right)$, ...

3) Sometimes, Always, or Never:

   a) The polar coordinates of a point are unique.

   Never.

   b) $(r, \theta) = (r, \theta + 2\pi)$.

   Always.

   c) $(r, \theta) = (-r, \theta + \pi)$.

   Always.

   d) $(r, \theta) = (-r, \theta)$.

   Sometimes. (True when $r = 0$).

4) Find polar coordinates for the point P with rectangular coordinates $P \colon \left(-3, 3\sqrt{3}\right)$.

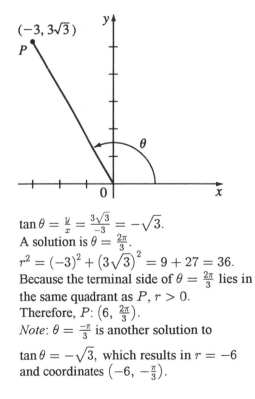

$\tan \theta = \frac{y}{x} = \frac{3\sqrt{3}}{-3} = -\sqrt{3}$.
A solution is $\theta = \frac{2\pi}{3}$.
$r^2 = (-3)^2 + \left(3\sqrt{3}\right)^2 = 9 + 27 = 36$.
Because the terminal side of $\theta = \frac{2\pi}{3}$ lies in the same quadrant as P, $r > 0$.
Therefore, $P \colon \left(6, \frac{2\pi}{3}\right)$.
Note: $\theta = \frac{-\pi}{3}$ is another solution to

$\tan \theta = -\sqrt{3}$, which results in $r = -6$ and coordinates $\left(-6, -\frac{\pi}{3}\right)$.

Areas and Lengths in Polar Coordinates

Concepts to Master

A. Area of a region described by polar equations
B. Length of a curve described by a polar equation

Summary and Focus Questions

A. The area of a region bounded by
$\theta = a$, $\theta = b$, $r = f(\theta)$ where f is
continuous and positive and
$0 \le b - a \le 2\pi$ is
$\int_a^b \frac{1}{2}[f(\theta)]^2 \, d\theta.$

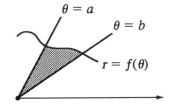

1) Find a definite integral for each region:

a) the shaded area

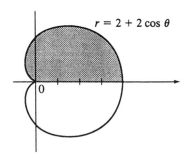

$r = 2 + 2\cos\theta$

$\text{Area} = \int_0^\pi \frac{1}{2}(2 + 2\cos\theta)^2 \, d\theta$
$= 2\int_0^\pi (1 + \cos\theta)^2 \, d\theta.$

b) the region bounded by $\theta = \frac{\pi}{3}$,
$\theta = \frac{\pi}{2}$, $r = e^\theta$.

$\text{Area} = \int_{\pi/3}^{\pi/2} \frac{1}{2}e^{2\theta} \, d\theta.$

c)   the shaded area

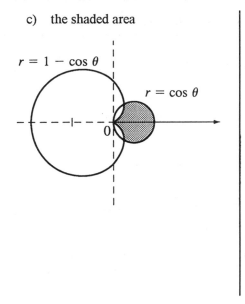

$r = 1 - \cos\theta$

$r = \cos\theta$

The area is between 2 curves so we must
first determine where the curves intersect:
$1 - \cos\theta = \cos\theta$, $1 = 2\cos\theta$,
$\cos\theta = \frac{1}{2}$. Therefore $\theta = \frac{\pi}{3}, -\frac{\pi}{3}$.
For $-\frac{\pi}{3} \leq \theta \leq \frac{\pi}{3}$,
$\cos\theta \geq 1 - \cos\theta$ so the area is

$= \int_{-\pi/3}^{\pi/3} \frac{1}{2}[\cos\theta]^2 \, d\theta$

$\quad - \int_{-\pi/3}^{\pi} \frac{1}{2}(1 - \cos\theta)^2 \, d\theta$

$= \int_{-\pi/3}^{\pi/3} \left(-\frac{1}{2} + \cos\theta\right) d\theta$

B. A curve given in polar coordinates by $r = f(\theta)$ for $a \leq \theta \leq b$ has arc length

$$\int_a^b \sqrt{[f(\theta)]^2 + [f'(\theta)]^2} \, d\theta.$$

2)   Set up a definite integral for the
length of each curve.

a)   the curve $r = e^{2\theta}$ for $0 \leq \theta \leq 1$.

$\int_0^1 \sqrt{(e^{2\theta})^2 + (2e^{2\theta})^2} \, d\theta = \sqrt{5}\int_0^1 e^{2\theta} \, d\theta.$

b)   the curve sketched below.

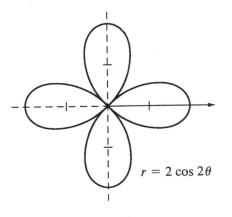

$r = 2\cos 2\theta$

Since the curve is quite symmetric, the arc
length is 8 times the length of half of one
leaf:

$8\int_0^{\pi/4} \sqrt{[2\cos 2\theta]^2 + [-4\sin 2\theta]^2} \, d\theta$

$= 8\int_0^{\pi/4} \sqrt{4\cos^2 2\theta + 16\sin^2 2\theta} \, d\theta$

$= 16\int_0^{\pi/4} \sqrt{1 + 3\sin^2 2\theta} \, d\theta$

Conic Sections in Polar Coordinates

Concepts to Master

Focus-directrix definitions of conics; Eccentricity; Equations of conics in polar form

---

Summary and Focus Questions

Conic sections, as you may recall, refer to parabolas, ellipses, and hyperbolas (curves obtained by intersecting a plane and a cone in various ways). Their equations and graphs in rectangular coordinates are given below.

Conic Section	Equation	Graph(s)
Parabola	$x^2 = 4py$	
Parabola	$y^2 = 4px$	

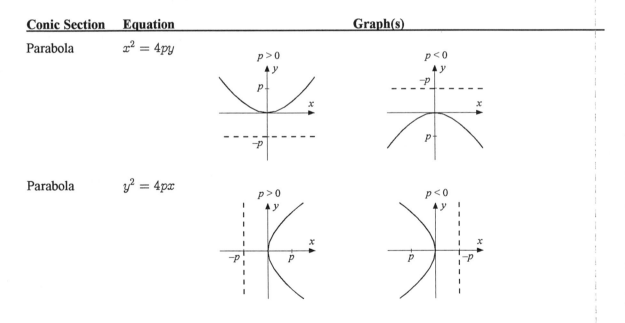

Ellipse $\qquad \dfrac{x^2}{a^2} + \dfrac{y^2}{b^2} = 1$

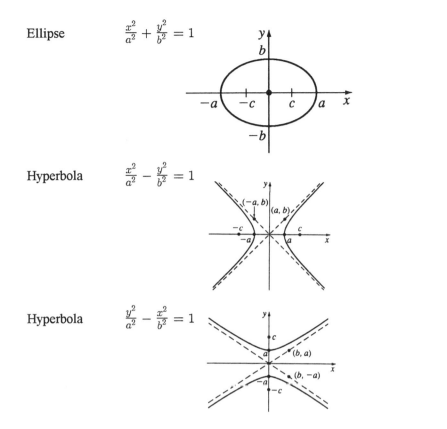

Hyperbola $\qquad \dfrac{x^2}{a^2} - \dfrac{y^2}{b^2} = 1$

Hyperbola $\qquad \dfrac{y^2}{a^2} - \dfrac{x^2}{b^2} = 1$

These conic sections may be described in a uniform fashion.

Let F be a fixed point (<u>focus</u>), l be a fixed line (<u>directrix</u>), and e be a positive constant* (<u>eccentricity</u>). The set of all points P such that $\dfrac{|PF|}{|Pl|} = e$ is a conic. The table on the next page describes the type of conics.

---

* This use of e for eccentricity is not to be confused with the use of e as the symbol for the base of natural logarithms.

b)   eccentricity $\frac{1}{2}$, directrix $y = -3$.

$e = \frac{1}{2}$ and $d = 3$, so $r = \dfrac{\frac{3}{2}}{1 - \frac{1}{2}\cos\theta}$,

$r = \dfrac{3}{2 - \cos\theta}$.

c)   directrix $x = 4$ and is a parabola.

$e = 1$ and $d = 4$, so $r = \dfrac{4}{1 + \cos\theta}$.

3)   Find the eccentricity and directrix and identify the conic give by $r = \dfrac{3}{4 - 5\cos\theta}$.

Divide numerator and denominator by 4.

$r = \dfrac{\frac{3}{4}}{1 - \frac{5}{4}\cos\theta}$, $e = \frac{5}{4}$. Since $ed = \frac{3}{4}$,

$d = \frac{3}{4} \cdot \frac{1}{e} = \frac{3}{4} \cdot \frac{4}{5} = \frac{3}{5}$.

The trigonometric term is $-\cos\theta$ so the directrix is $x = -\frac{3}{5}$. Since $e > 1$, the conic is a hyperbola.

4)   What polar form does the equation of the graph below have?

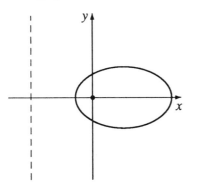

The graph is that of an ellipse. The directrix has the form $x = -d$ so the equation is $r = \dfrac{ed}{1 - e\cos\theta}$.

10

Infinite Sequences and Series

"I'M BEGINNING TO UNDERSTAND ETERNITY, BUT INFINITY IS STILL BEYOND ME."

Cartoons courtesy of Sidney Harris. Used by permission.

Sequences

Concepts to Master

A. Sequences; Limit of a sequence; Convergent; Divergent

B. Monotone sequences (increasing, decreasing); Bounded sequences; Sequences defined recursively

Summary and Focus Questions

A. A <u>sequence</u> is a list of numbers arranged in order:

$$\{a_n\} = a_1, a_2, a_3, \ldots, a_n, a_{n+1}, \ldots$$

a_n is the <u>nth term</u> of the sequence.

$\lim\limits_{n\to\infty} a_n = L$ means the sequence $\{a_n\}$ <u>converges to L</u>, that is, as n grows larger the a_n values get closer and closer to L. Formally, $\lim\limits_{n\to\infty} a_n = L$ means for all $\varepsilon > 0$ there exists a positive integer N such that $|a_n - L| < \varepsilon$ for all $n > N$. If the limit does not exist, $\{a_n\}$ <u>diverges</u>.

$\lim\limits_{n\to\infty} a_n = \infty$ means the a_n terms grow without bound.

One way to evaluate $\lim\limits_{n\to\infty} a_n$ is to find a real function $f(x)$ such that $f(n) = a_n$ for all n. If $\lim\limits_{n\to\infty} f(x) = L$, then $\lim\limits_{n\to\infty} a_n = L$. The converse is false.

All limit laws for limits of functions at infinity are valid for convergent sequences. A version of the Squeeze Theorem also holds:

If $a_n \leq b_n \leq c_n$ for all $n \geq N$ and $\lim\limits_{n\to\infty} a_n = \lim\limits_{n\to\infty} c_n = L$, then $\lim\limits_{n\to\infty} b_n = L$.

1) Find the fourth term of the sequence
$a_n = \frac{(-1)^n}{n^2}$.

$a_4 = \frac{(-1)^4}{4^2} = \frac{1}{16}$.

2) Determine whether each converges.

   a) $a_n = \frac{n}{n^2+1}$.

$\lim\limits_{n\to\infty} \frac{n}{n^2+1} = $ (divide by n^2)

$$\lim\limits_{n\to\infty} \frac{\frac{1}{n}}{1+\frac{1}{n^2}} = \frac{0}{1+0} = 0.$$

Thus a_n converges to 0.

   b) $b_n = \frac{n^2+1}{2n}$.

b_n grows without bound so $\{b_n\}$ diverges.
We may write $\lim\limits_{n\to\infty} \frac{n^2+1}{2n} = \infty$.

   c) $c_n = \frac{(-1)^n n}{n+1}$.

The sequence $\frac{n}{n+1}$ converges to 1 so
$c_n = \frac{(-1)^n n}{n+1}$ is alternating values near 1
and -1. Thus $\{c_n\}$ diverges.

3) Find $\lim\limits_{n\to\infty} \frac{\cos n}{n}$.

0. Since $-1 \le \cos n \le 1$, $-\frac{1}{n} \le \frac{\cos n}{n} \le \frac{1}{n}$.
Thus as $n \to \infty$, $\frac{\cos n}{n} \to 0$.

4) Find $\lim\limits_{n\to\infty} \frac{\ln n}{n}$.

Let $f(x) = \frac{\ln x}{x}$.

$\lim\limits_{x\to\infty} \frac{\ln x}{x} \left(= \frac{0}{0} \text{ form} \right)$

$= \lim\limits_{x\to\infty} \frac{1/x}{1}$ (L'Hôpital's Rule) $= 0$.

Thus $\lim\limits_{x\to\infty} \frac{\ln n}{n} = 0$.

B. $\{a_n\}$ is <u>increasing</u> means $a_{n+1} \ge a_n$ for all n.
   $\{a_n\}$ is <u>decreasing</u> means $a_{n+1} \le a_n$ for all n.
   $\{a_n\}$ is <u>monotonic</u> if it is either increasing or decreasing.
   One way to show $\{a_n\}$ is increasing is to show that $\frac{da_n}{dn} \ge 0$ (treating n as a
   real number).
   $\{a_n\}$ is <u>bounded above</u> means $a_n \le M$ for some M and all n.
   $\{a_n\}$ is <u>bounded below</u> means $a_n \ge m$ for some m and all n.
   $\{a_n\}$ is <u>bounded</u> if it is both bounded above and bounded below.

Series

Concepts to Master

A. Infinite series; Partial sums; Convergent and divergent series; Convergent series laws

B. Geometric series; Value of a converging geometric series; Harmonic series

Summary and Focus Questions

A. Adding up all the terms of a sequence $\{a_n\}$ is an <u>(infinite) series</u>:

$$\sum_{k=1}^{\infty} a_k = a_1 + a_2 + a_3 + \ldots + a_n + \ldots$$

If we stop adding after n terms, we have the <u>nth partial sum</u> of the series:

$$s_n = \sum_{k=1}^{\infty} a_k = a_1 + a_2 + a_3 + \ldots + a_n + \ldots$$

$\sum_{n=1}^{\infty} a_n$ <u>converges (to s)</u> means $\lim\limits_{n \to \infty} s_n$ exists (and is s). Thus a series converges if the limit of its sequence of partial sums exists.

If that limit does not exist, then $\sum_{n=1}^{\infty} a_n$ <u>diverges</u>.

If $\sum_{n=1}^{\infty} a_n$ converges, then $\lim\limits_{n \to \infty} a_n = 0$. Rewritten in a contrapositive form, this is the <u>Test for Divergence</u>:

If $\lim\limits_{n \to \infty} a_n \neq 0$, $\sum_{k=1}^{\infty} a_k$ diverges.

The converse is false - just because the terms a_n get small is not enough to conclude the series converges.

If $\displaystyle\sum_{n=1}^{\infty} a_n$ converges (to L) and $\displaystyle\sum_{n=1}^{\infty} b_n$ converges (to M) then:

$$\sum_{n=1}^{\infty} (a_n \pm b_n) \text{ converges (to } L \pm M) .$$

$$\sum_{n=1}^{\infty} c a_n \text{ converges (to } cL) \text{ for } c \text{ any constant.}$$

1) A series converges if the _____ of the sequence of _____ exists.

limit, partial sums

2) Find the first four partial sums of $\displaystyle\sum_{k=1}^{\infty} \frac{1}{k^2}$.

$s_1 = \frac{1}{1^2} = 1.$

$s_2 = \frac{1}{1^2} + \frac{1}{2^2} = 1 + \frac{1}{4} = \frac{5}{4}.$

$s_3 = \frac{1}{1^2} + \frac{1}{2^2} + \frac{1}{3^2} = \frac{5}{4} + \frac{1}{9} = \frac{49}{36}.$

$s_4 = \frac{1}{1^2} + \frac{1}{2^2} + \frac{1}{3^2} + \frac{1}{4^2} = \frac{49}{36} + \frac{1}{16} = \frac{205}{144}.$

3) Sometimes, Always, or Never:
If $\displaystyle\lim_{n \to \infty} a_n = 0$, $\displaystyle\sum_{n=1}^{\infty} a_n$ converges.

Sometimes.

4) Suppose $\displaystyle\sum_{n=1}^{\infty} a_n = 3$ and $\displaystyle\sum_{n=1}^{\infty} b_n = 4$.
Evaluate each of the following:

a) $\displaystyle\sum_{n=1}^{\infty} (a_n + 2b_n)$

$3 + 2(4) = 11.$

b) $\displaystyle\sum_{n=1}^{\infty} \frac{a_n}{5}$

$\frac{3}{5}.$

c) $\displaystyle\sum_{n=1}^{\infty} \frac{1}{a_n}$

This diverges because if $\displaystyle\sum_{n=1}^{\infty} a_n = 3$, then
$\displaystyle\lim_{n\to\infty} a_n = 0$. Thus $\displaystyle\lim_{n\to\infty} \frac{1}{a_n} \neq 0$, so $\displaystyle\sum_{n=1}^{\infty} \frac{1}{a_n}$
diverges.

5) Sometimes, Always, or Never:

a) $\sum_{n=1}^{\infty} a_n$ converges, but $\lim_{n\to\infty} a_n$ does
not exist.

Never. If $\sum_{n=1}^{\infty} a_n$ converges, then $\lim_{n\to\infty} a_n$ exists and is 0.

b) $\sum_{n=1}^{\infty} a_n$ converges, $\sum_{n=1}^{\infty} (a_n + b_n)$
converges, but $\sum_{n=1}^{\infty} b_n$ diverges.

Never. $\sum_{n=1}^{\infty} b_n = \sum_{n=1}^{\infty} (a_n + b_n) - \sum_{n=1}^{\infty} a_n$ is
the difference of two convergent series and must converge.

6) Does $\sum_{n=1}^{\infty} \cos\left(\frac{1}{n}\right)$ converge?

No, the nth term does not approach 0.

B. A _geometric series_ has the form
$$\sum_{n=1}^{\infty} ar^{n-1} = a + ar + ar^2 + ar^3 + \dots$$
and converges (for $a \neq 0$) to $\frac{a}{1-r}$ if and only if $-1 < r < 1$.

The _harmonic series_ $\sum_{n=1}^{\infty} \frac{1}{n} = 1 + \frac{1}{2} + \frac{1}{3} + \dots + \frac{1}{n} + \dots$ diverges.

7) a) $\sum_{n=1}^{\infty} \frac{2^n}{100}$ is a geometric series in
which $a = $ _____ and $r = $ _____.

$\sum_{n=1}^{\infty} \frac{2^n}{100} = \frac{2}{100} + \frac{4}{100} + \frac{8}{100} + \dots$
so $a = \frac{2}{100}$ and $r = 2$.

b) Does it converge?

No, since $r \geq 1$.

8) $\sum_{n=1}^{\infty} \left(\frac{-2}{9}\right)^n = $ _____.

A geometric series with $a = r = -\frac{2}{9}$
which converges to $\frac{a}{1-r} = -\frac{2}{11}$.

9) Does $\sum_{n=1}^{\infty} \frac{6}{n}$ converge?

It diverges, since it is a multiple of the harmonic series.

10) $\sum_{n=1}^{\infty} \frac{2^n}{3^{n+1}} =$ _____.

$\sum_{n=1}^{\infty} \frac{2^n}{3^{n+1}} = \sum_{n=1}^{\infty} \frac{2}{9}\left(\frac{2}{3}\right)^{n-1}$, which is a

geometric series with $a = \frac{2}{9}$, $r = \frac{2}{3}$ that

converges to $\frac{\frac{2}{9}}{1-\frac{2}{3}} = \frac{2}{3}$.

11) Achilles gives the tortoise a 100 m head start. If Achilles runs at 5 m/sec and the tortoise at $\frac{1}{2}$ m/sec how far has the tortoise traveled by the time Achilles catches him?

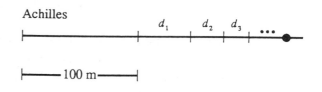

Let $d_1 =$ distance tortoise traveled while Achilles was running to the tortoise's starting point.

$d_1 = \left(\frac{100\,\text{m}}{5\,\text{m/sec}}\right)\left(\frac{1}{2}\,\text{m/sec}\right) = 10\,\text{m}.$

Let $d_2 =$ distance tortoise traveled while Achilles was running the distance d_1. Since $d_1 = 10$ and Achilles runs at 5 m/sec $d_2 = \left(\frac{10\,\text{m}}{5\,\text{m/sec}}\right)\frac{1}{2}\,\text{m/sec} = 1\,\text{m}.$

For each n, $d_n = \left(\frac{d_{n-1}}{5}\right)\frac{1}{2} = \frac{d_{n-1}}{10}.$
The total distance traveled by the tortoise is $\sum_{n=1}^{\infty} d_n = 10 + 1 + \frac{1}{10} + \ldots$ which is a geometric series with $a = 10$, $r = \frac{1}{10}$. This converges to $\frac{10}{1-\frac{1}{10}} \doteq \frac{100}{9}$ m.

The Integral Test

Concepts to Master

A. Integral Test for convergence; p-series
B. Estimate the sum of a convergent series

Summary and Focus Questions

This section and sections 10.4, 10.5, and 10.6 provide tests to determine whether a series converges. The tests will not give a value for the series, only whether there is such a value, and in a few cases a good estimate if the value exists.

A. <u>Integral Test</u>: Let $\{a_n\}$ be a sequence. Suppose there is $f(x)$, a positive, continuous, and decreasing function on $[1, \infty)$ such that $a_n = f(n)$ for all n. Then $\sum\limits_{n=1}^{\infty} a_n$ converges if and only if $\int_1^{\infty} f(x)dx$ converges.

The Integral Test works when a_n has the form of a function whose antiderivative is easily found. It is one of the few general tests that gives necessary and sufficient conditions for convergence.

A <u>p-series</u> has the form $\sum\limits_{n=1}^{\infty} \frac{1}{n^p}$, ($p$ a constant), and converges (by the Integral Test) if and only if $p > 1$.

1) True or False:
   The Integral Test will determine to what value a series converges.

   False. The test only indicates whether a series converges.

2) Test $\sum_{n=1}^{\infty} \frac{n}{e^n}$ for convergence.

$\frac{n}{e^n}$ suggests the function $f(x) = xe^{-x}$ when on $[1, \infty)$ is continuous and decreasing. $\int_1^{\infty} xe^{-x} = \lim_{t\to\infty} \int_1^t xe^{-x} \, dx$.

Using integration by parts,
$$\begin{pmatrix} u = x & dv = e^{-x} \, dx \\ du = dx & dv = -e^{-x} \end{pmatrix}$$
$\int_1^t xe^{-x} \, dx = -xe^{-x} - e^{-x} \Big|_1^t = \frac{2}{e} - \frac{t+1}{e^t}$.

$\lim_{t\to\infty} \left(\frac{2}{e} - \frac{t+1}{e^t} \right) = \frac{2}{e} - 0 = \frac{2}{e}$.

$\left(\text{By L'Hôpital's Rule, } \frac{t+1}{e^t} \to 0. \right)$

Thus $\int_1^{\infty} xe^{-x} \, dx = \frac{2}{e}$, so $\sum_{n=1}^{\infty} \frac{n}{e^n}$ converges.

Note: We can<u>not</u> conclude than $\sum_{n=1}^{\infty} \frac{n}{e^n} = \frac{2}{e}$.

3) Which of these converge?

a) $\sum_{n=1}^{\infty} \frac{1}{n^2}$.

Converges (p-series with $p = 2$).

b) $\sum_{n=1}^{\infty} \frac{1}{\sqrt{n}}$.

Diverges (p-series with $p = \frac{1}{2}$).

c) $\sum_{n=1}^{\infty} \frac{3}{2n^3}$.

Converges. $\sum_{n=1}^{\infty} \frac{3}{2n^3} = \frac{3}{2} \sum_{n=1}^{\infty} \frac{1}{n^3}$ is a multiple of a p-series with $p = 3$.

d) $\sum_{n=1}^{\infty} \left(\frac{1}{n^3} + \frac{1}{8^n} \right)$.

Converges. $\sum_{n=1}^{\infty} \left(\frac{1}{n^3} + \frac{1}{8^n} \right) = \sum_{n=1}^{\infty} \frac{1}{n^3} + \sum_{n=1}^{\infty} \frac{1}{8^n}$, the sum of a convergent p-series ($p = 3$) and a convergent geometric series $\left(r = \frac{1}{8} \right)$.

B. If $\sum a_n = s$ is determined to converge by the Interal Test using $f(x)$ then the error $R_n = s - s_n$ between the series value and the nth partial sum satisfies
$$\int_{n+1}^{\infty} f(x)dx \le s - s_n \le \int_n^{\infty} f(x)dx.$$

4) a) Find the sixth partial sum of $\sum_{n=1}^{\infty} \frac{1}{n^2}$.

$s_6 = 1 + \frac{1}{4} + \frac{1}{9} + \frac{1}{16} + \frac{1}{25} + \frac{1}{36}$

$= \frac{5369}{3600} \approx 1.491$.

b) Estimate the difference between your answer to part a) and the exact value of $s = \sum\limits_{n-1}^{\infty} \frac{1}{n^2}$.

We know by the Integral Test using $f(x) = \frac{1}{x^2}$ that $\sum\limits_{n=1}^{\infty} \frac{1}{n^2}$ converges. Thus
$\int_7^{\infty} \frac{1}{x^2}\, dx \leq s - s_6 \leq \int_6^{\infty} \frac{1}{x^2}\, dx.$

$\int_7^{\infty} \frac{1}{x^2}\, dx = \lim\limits_{t\to\infty} \int_7^t x^{-2}\, dx$
$\qquad = \lim\limits_{t\to\infty} \left(\frac{1}{7} - \frac{1}{t}\right) = \frac{1}{7}.$

Likewise $\int_6^{\infty} \frac{1}{x^2}\, dx = \frac{1}{6}.$

Therefore $\frac{1}{7} \leq s - s_6 \leq \frac{1}{6}.$

c) Estimate $s = \sum\limits_{n=1}^{\infty} \frac{1}{n^2}$ using the results of parts a) and b).

$\frac{1}{7} \leq s - s_6 \leq \frac{1}{6}.$

$\frac{1}{7} + s_6 \leq s \leq \frac{1}{6} + s_6.$

$\frac{1}{7} + \frac{5369}{3600} \leq s \leq \frac{1}{6} + \frac{5369}{3600}$

$\frac{41183}{25200} \leq s \leq \frac{5969}{3600}$

$1.634 \leq s \leq 1.658$

It turns out that $s = \frac{\pi^2}{6} \approx 1.645$, so s_6 is a rather good estimate.

The Comparison Test

Concepts to Master

A. Comparison Test; Limit Comparison Test
B. Estimate the sum of a convergent series

Summary and Focus Questions

A. The following tests may be used to determine the convergence of a series whose *terms are all positive*.

<u>Comparison Test</u>:

1. If there is a convergent series $\sum_{n=1}^{\infty} b_n$ such that $0 \leq a_n \leq b_n$ for all $n \geq N$, then $\sum_{n=1}^{\infty} a_n$ converges.

2. If there is a divergent series $\sum_{n=1}^{\infty} b_n$ such that $0 \leq b_n \leq a_n$ for all $n \geq N$, then $\sum_{n=1}^{\infty} a_n$ diverges.

<u>Limit Comparison Test</u>:

Let $\sum_{n=1}^{\infty} a_n$ and $\sum_{n=1}^{\infty} b_n$ be positive term series.

1. If $\lim_{n \to \infty} \frac{a_n}{b_n} = c > 0$, then $\sum_{n=1}^{\infty} a_n$ and $\sum_{n=1}^{\infty} b_n$ either both converge or both diverge.

2. If $\lim\limits_{n\to\infty} \frac{a_n}{b_n} = 0$, and $\sum\limits_{n=1}^{\infty} b_n$ converges, then $\sum\limits_{n=1}^{\infty} a_n$ converges.

3. If $\lim\limits_{n\to\infty} \frac{a_n}{b_n} = 0$, and $\sum\limits_{n=1}^{\infty} b_n$ diverges, then $\sum\limits_{n=1}^{\infty} a_n$ diverges.

To successfully use either test on a series $\sum\limits_{n=1}^{\infty} a_n$, you must come up with another series $\sum\limits_{n=1}^{\infty} b_n$ (geometric, p-series, ...) that you know is convergent or divergent and compare a_n to b_n.

1) Determine whether or not each of the following converge. Find a series $\sum\limits_{n=1}^{\infty} b_n$ to use with the Comparison Test or Limit Comparison Test.

a) $\sum\limits_{n=1}^{\infty} \frac{1}{n^2+2n}$,

$b_n = $ _____ .

You must have a hunch beforehand whether $\sum\limits_{n=1}^{\infty} \frac{1}{n^2+2n}$ converges. Because

$\frac{1}{n^2+2n}$ is "like" $\frac{1}{n^2}$ for large n, and $\sum\limits_{n=1}^{\infty} \frac{1}{n^2}$ is a converging p-series, our hunch is the given series converges. Now we have an idea of what kind of b_n to look for. Use $b_n = \frac{1}{n^2}$. Since $\frac{1}{n^2+2n} < \frac{1}{n^2}$ and $\sum\limits_{n=1}^{\infty} \frac{1}{n^2}$

converges, $\sum\limits_{n=1}^{\infty} \frac{1}{n^2+2n}$ converges by the Comparison Test.

b) $\sum\limits_{n=1}^{\infty} \frac{\sqrt[3]{n}}{n+4}$,

$b_n = $ _____ .

Your hunch should be $\frac{\sqrt[3]{n}}{n+4}$ is "like"
$\frac{\sqrt[3]{n}}{n} = \frac{1}{\sqrt[3]{n^2}}$. Use $b_n = \frac{1}{\sqrt[3]{n^2}}$. Then

$$\lim_{n\to\infty} \frac{a_n}{b_n} = \lim_{n\to\infty} \frac{\frac{\sqrt[3]{n}}{n+4}}{\frac{1}{\sqrt[3]{n^2}}} = \lim_{n\to\infty} \frac{n}{n+4} = 1.$$

Since $\sum\limits_{n=1}^{\infty} \frac{1}{\sqrt[3]{n^2}}$ diverges $\left(\text{a } p\text{-series with}\right.$

$p = \frac{2}{3}\big)$, $\sum\limits_{n=1}^{\infty} \frac{\sqrt[3]{n}}{n+4}$ diverges by the Limit

Comparison Test.

c) $\sum\limits_{n=1}^{\infty} \frac{1}{n+2^n}$,

  $b_n =$ _____.

Use $b_n = \frac{1}{2^n}$. Then for $n \geq 1$, $\frac{1}{n+2^n} \leq \frac{1}{2^n}$.

Since $\sum\limits_{n=1}^{\infty} \frac{1}{2^n}$ converges $\left(\text{a geometric series}\right.$

with $r = \frac{1}{2}\big)$, $\sum\limits_{n=1}^{\infty} \frac{1}{n+2^n}$ converges.

2) Suppose $\{a_n\}$ and $\{b_n\}$ are positive sequences. Sometimes, Always, Never:

a)   If $\lim\limits_{n\to\infty} \frac{a_n}{b_n} = 0$ and $\sum\limits_{k=1}^{\infty} a_k$ converges,

   then $\sum\limits_{k=1}^{\infty} b_k$ converges.

Sometimes. For $a_n = \frac{1}{n^3}$ and $b_n = \frac{1}{n^2}$ it is true. For $a_n = \frac{1}{n^2}$ and $b_n = \frac{1}{n}$ it is false.

b)   If $\lim\limits_{n\to\infty} \frac{a_n}{b_n} = \infty$ and $\sum\limits_{k=1}^{\infty} b_n$ diverges,

   then $\sum\limits_{k=1}^{\infty} a_n$ diverges.

Always.

B. If $\sum\limits_{n=1}^{\infty} a_n = s$ converges by the comparison Test using $t = \sum\limits_{n=1}^{\infty} b_n$, then

$$s - s_n \leq t - t_n.$$

This means that $t - t_n$ (which may be easier to calculate) is an upper estimate for $s - s_n$.

3)   In question 1c), $\sum\limits_{n=1}^{\infty} \frac{1}{n+2^n}$ converges.

a)   Find s_4, the fourth partial sum.

$$s_4 = \frac{1}{1+2} + \frac{1}{2+4} + \frac{1}{3+8} + \frac{1}{4+16}$$

$$= \frac{1}{3} + \frac{1}{6} + \frac{1}{11} + \frac{1}{20} = \frac{141}{220}.$$

b) Estimate the difference between
    this series and its fourth partial
    sum.

From 1c), $b_n = \frac{1}{2^n}$ may be used to show
$\sum \frac{1}{n+2^n}$ converges. Let $s = \sum\limits_{n=1}^{\infty} \frac{1}{n+2^n}$ and
$t = \sum\limits_{n=1}^{\infty} \frac{1}{2^n}$. Then $s - s_4 \leq t - t_4$.
Since t is the result of a geometric series
we can calculate it:
$$t = \frac{\frac{1}{2}}{1-\frac{1}{2}} = 1$$
$$t_4 = \frac{1}{2} + \frac{1}{4} + \frac{1}{8} + \frac{1}{16} = \frac{31}{32}.$$
Thus $t - t_4 = 1 - \frac{31}{32} = \frac{1}{32}$ and
$s - s_4 \leq \frac{1}{32}$. Therefore we know $s_4\left(\frac{141}{220}\right)$ is
within $\frac{1}{32}$ of the value of s.

Alternating Series

Concepts to Master

A. Alternating Series Tests
B. Estimating the sum of a convergent alternating series

Summary and Focus Questions

A. An <u>alternating series</u> has successive terms of opposite signs - that is, has either

the form $\sum_{n=1}^{\infty}(-1)^n a_n$ or $\sum_{n=1}^{\infty}(-1)^{n+1}a_n$, where $a_n > 0$.

The <u>Alternating Series Test</u>:

    If $\{a_n\}$ is a decreasing sequence with $\lim_{n\to\infty} a_n = 0$, then $\sum_{n=1}^{\infty}(-1)^n a_n$

    converges.

1) Is $\sum_{n=1}^{\infty}\frac{\sin n}{n}$ an alternating series?

    No, although some terms are positive and others negative.

2) Determine whether each converge:

    a) $\sum_{n=1}^{\infty}\frac{(-1)^n}{e^n}$.

    This is an alternating series with $a_n = \frac{1}{e^n}$.
    $\frac{1}{e^{n+1}} \leq \frac{1}{e^n}$, so the terms decrease.
    Since $\lim_{n\to\infty}\frac{1}{e^n} = 0$, $\sum_{n=1}^{\infty}\frac{(-1)^n}{e^n}$ converges by the
    Alternating Series Test.

    b) $\sum_{n=1}^{\infty}\frac{(-1)^n n}{n+1}$.

    This is an alternating series but the other conditions for the test do not hold.
    However, $\lim_{n\to\infty}\frac{(-1)^n n}{n+1} \neq 0$, so $\sum_{n=1}^{\infty}\frac{(-1)^n n}{n+1}$
    diverges by the Divergence Test.

B. If $\sum\limits_{n=1}^{\infty}(-1)^n a_n$ is an alternating series with $0 \leq a_{n+1} \leq a_n$ and $\lim\limits_{n\to\infty} a_n = 0$ then

$$|s_n - s| \leq a_{n+1}.$$

Since the difference between the limit of the series and the nth partial sum does not exceed a_{n+1}, s_n may be used to approximate s to within an accuracy of a_{n+1} by s_n.

3) The series $\sum\limits_{n=1}^{\infty}\frac{(-1)^n}{\sqrt{n+1}}$ converges.

Estimate the error between the sum of the series and its fifteenth partial sum.

Since the series alternates and $\frac{1}{\sqrt{n+1}}$ decreases to 0, the error $|s_{15} - s|$ does not exceed $a_{16} = \frac{1}{\sqrt{16+1}} = 0.2$.

4) For what value of n is s_n, the nth partial sum, within 0.001 of

$$s = \sum\limits_{n=1}^{\infty}\frac{(-1)^{n+1}}{2n+5}?$$

$|s_n - s| \leq a_{n+1} = \frac{1}{(2n+1)+5} \leq 0.001$
$2n + 6 \geq 1000$, $2n \geq 994$, $n \geq 497$.
Let $n = 497$. Then s_{497} is within 0.001 of s.

5) Approximate $\sum\limits_{n=1}^{\infty}\frac{(-1)^{n+1}}{n^3}$ to within 0.005.

$|s_n - s| \leq a_{n+1} = \frac{1}{(n+1)^3} \leq 0.005$
$(n + 1)^3 \geq \frac{1}{0.005} = 200$.
For $n = 5$, $(n + 1)^3 = 6^3 = 216 > 200$.
Therefore s_5 is within 0.005 of s.
$s_5 = \frac{1}{1^3} - \frac{1}{2^3} + \frac{1}{3^3} - \frac{1}{4^3} + \frac{1}{5^3}$
$\quad = 1 - \frac{1}{8} + \frac{1}{27} - \frac{1}{64} + \frac{1}{125} \approx 0.9044$.
We do not know to what the series converges, but that value is within 0.005 of 0.9044.

Absolute Convergence and the Ratio and Root Tests

Concepts to Master

A. Absolute convergence; Conditional convergence
B. Ratio Test; Root Test

Summary and Focus Questions

A. For any series (with a_n not necessarily positive or alternating) the concept of convergence may be split into two subconcepts:

$$\sum_{k=1}^{\infty} a_k \underline{\text{ converges absolutely}} \text{ means that } \sum_{k=1}^{\infty} |a_k| \text{ converges.}$$

$$\sum_{k=1}^{\infty} a_k \underline{\text{ converges conditionally}} \text{ means that } \sum_{k=1}^{\infty} a_k \text{ converges and } \sum_{k=1}^{\infty} |a_k|$$
diverges.

Either one of absolute convergence or conditional convergence implies (ordinary) convergence. Conversely, convergence implies either absolute or conditional convergence. Thus *every* series must behave in exactly one of these three ways: diverge, converge absolutely, or converge conditionally.

1) Sometimes, Always, or Never:

   a) If $\sum_{n=1}^{\infty} |a_n|$ diverges, then $\sum_{n=1}^{\infty} a_n$ diverges.

   Sometimes. True for $a_n = n$ but false for $a_n = \frac{(-1)^n}{n}$.

267

b) If $\sum\limits_{n=1}^{\infty} a_n$ diverges, then

$\sum\limits_{n=1}^{\infty} |a_n|$ diverges.

Always.

c) If $a_n \geq 0$ for all n, then

$\sum\limits_{n=1}^{\infty} a_n$ is not conditionally

convergent.

Always.

2) Determine whether each converges
conditionally, converges absolutely, or
diverges.

a) $\sum\limits_{n=1}^{\infty} \frac{(-1)^n}{\sqrt{n}}$.

The series is alternating with $\frac{1}{\sqrt{n}}$
decreasing to 0 so it converges. It remains
to check for absolute convergence:

$\sum\limits_{n=1}^{\infty} \left| \frac{(-1)^n}{\sqrt{n}} \right| = \sum\limits_{n=1}^{\infty} \frac{1}{\sqrt{n}}$ diverges $\left(p\text{-series with} \right.$

$\left. p = \frac{1}{2} \right)$. Thus $\sum\limits_{n=1}^{\infty} \frac{(-1)^n}{\sqrt{n}}$ converges

conditionally.

b) $\sum\limits_{n=1}^{\infty} \frac{\sin n + \cos n}{n^3}$.

Check for absolute convergence first:

$\sum\limits_{n=1}^{\infty} \left| \frac{\sin n + \cos n}{n^3} \right| = \sum\limits_{n=1}^{\infty} \frac{|\sin n + \cos n|}{n^3}$.

Since $\left| \sin n + \cos n \right| \leq 2$ and $\sum\limits_{n=1}^{\infty} \frac{2}{n^3}$
converges (it is a p-series with $p = 3$),

$\sum\limits_{n=1}^{\infty} \frac{|\sin n + \cos n|}{n^3}$ converges by the Comparison

Test. Thus $\sum\limits_{n=1}^{\infty} \frac{\sin n + \cos n}{n^3}$ converges

absolutely.

3) Does $\sum\limits_{n=1}^{\infty} \frac{\sin e^n}{e^n}$ converge?

Yes. The series is not alternating but does contain both positive and negative terms. Check for absolute convergence:
$$\sum_{n=1}^{\infty} \left| \frac{\sin e^n}{e^n} \right| = \sum_{n=1}^{\infty} \frac{|\sin e^n|}{e^n}.$$
Since $\left| \sin e^n \right| \le 1$, $\frac{|\sin e^n|}{e^n} \le \frac{1}{e^n}$.
$\sum\limits_{n=1}^{\infty} \frac{1}{e^n}$ converges $\Big($ geometric series with $r = \frac{1}{e} \Big)$ so by the Comparison Test $\sum\limits_{n=1}^{\infty} \left| \frac{\sin e^n}{e^n} \right|$ converges. Therefore $\sum\limits_{n=1}^{\infty} \frac{\sin e^n}{e^n}$ converges absolutely and hence converges.

B. Let a_n be a sequence of nonzero terms. The following may be used to determine whether $\sum\limits_{n=1}^{\infty} a_n$ converges.

<u>Ratio Test</u>:

1. If $\lim\limits_{n\to\infty} \left| \frac{a_{n+1}}{a_n} \right| = L < 1$, then $\sum\limits_{n=1}^{\infty} a_n$ converges absolutely (and therefore converges).

2. If $\lim\limits_{n\to\infty} \left| \frac{a_{n+1}}{a_n} \right| = L > 1$ or $\lim\limits_{n\to\infty} \left| \frac{a_{n+1}}{a_n} \right| = \infty$, then $\sum\limits_{n=1}^{\infty} a_n$ diverges.

In case $\lim\limits_{n\to\infty} \left| \frac{a_{n+1}}{a_n} \right| = 1$, the Ratio Test fails - $\sum\limits_{n=1}^{\infty} a_n$ may converge or diverge.

<u>Root Test</u>:

1. If $\lim\limits_{n\to\infty} \sqrt[n]{|a_n|} = L < 1$, then $\sum\limits_{n=1}^{\infty} a_n$ converges absolutely (and therefore converges).

2. If $\lim\limits_{n\to\infty} \sqrt[n]{|a_n|} = L > 1$ or $\lim\limits_{n\to\infty} \sqrt[n]{|a_n|} = \infty$, then $\sum\limits_{n=1}^{\infty} a_n$ diverges.

If $\lim\limits_{n\to\infty} \sqrt[n]{|a_n|} = 1$, the Root Test fails - $\sum\limits_{n=1}^{\infty} a_n$ may converge or diverge.

The Ratio and Root Tests involve no other series or functions and so are relatively easy to use, but both fail for $L = 1$. The Ratio Test usually provides an answer when a_n contains exponentials and/or factorials or when a_n is defined by a recurrence relation. It will not work when a_n is a rational function of n. The Root Test works best when a_n contains an expression to the nth power.

4) Determine whether each converges by the Ratio Test:

a) $\displaystyle\sum_{n=1}^{\infty} \frac{n}{4^n}$.

$$\lim_{n\to\infty} \left|\frac{a_{n+1}}{a_n}\right| = \lim_{n\to\infty} \frac{\frac{n+1}{4^{n+1}}}{\frac{n}{4^n}} = \lim_{n\to\infty} \frac{n+1}{4n} = \frac{1}{4}.$$

Thus $\displaystyle\sum_{n=1}^{\infty} \frac{n}{4^n}$ converges absolutely and therefore converges.

b) $\displaystyle\sum_{n=1}^{\infty} \frac{(-4)^n}{n!}$.

$$\lim_{n\to\infty} \left|\frac{a_{n+1}}{a_n}\right| = \lim_{n\to\infty} \left|\frac{\frac{(-4)^{n+1}}{(n+1)!}}{\frac{(-4)^n}{n!}}\right| = \lim_{n\to\infty} \frac{4}{n+1} = 0.$$

Thus $\displaystyle\sum_{n=1}^{\infty} \frac{(-4)^n}{n!}$ converges absolutely and therefore converges.

c) $\displaystyle\sum_{n=1}^{\infty} \frac{(-1)^n}{\sqrt[3]{n}}$.

$$\lim_{n\to\infty} \left|\frac{a_{n+1}}{a_n}\right| = \lim_{n\to\infty} \left|\frac{\frac{(-1)^{n+1}}{\sqrt[3]{n+1}}}{\frac{(-1)^n}{\sqrt[3]{n}}}\right| = \lim_{n\to\infty} \sqrt[3]{\frac{n}{n+1}} = 1.$$

The Ratio Test <u>fails</u>. The Alternating Series Test can be used to show that this series converges.

d) $\displaystyle\sum_{n=1}^{\infty} a_n$, where $a_1 = 4$ and $a_{n+1} = \frac{3a_n}{2n+1}$.

$$\lim_{n\to\infty} \left|\frac{a_{n+1}}{a_n}\right| = \lim_{n\to\infty} \frac{\frac{3a_n}{2n+1}}{a_n} = \lim_{n\to\infty} \frac{3}{2n+1} = 0.$$

Thus $\displaystyle\sum_{n=1}^{\infty} a_n$ converges absolutely and therefore converges.

5) Determine whether each converges by the Root Test:

a) $\sum\limits_{n=1}^{\infty} \frac{1}{(n+1)^n}$.

$$\lim_{n\to\infty} \sqrt[n]{|a_n|} = \lim_{n\to\infty} \sqrt[n]{\frac{1}{(n+1)^n}} = \lim_{n\to\infty} \frac{1}{n+1} = 0.$$

Thus $\sum\limits_{n=1}^{\infty} \frac{1}{(n+1)^n}$ converges absolutely and therefore converges.

b) $\sum\limits_{n=1}^{\infty} \frac{3^n}{n^3}$.

$$\lim_{n\to\infty} \sqrt[n]{|a_n|} = \lim_{n\to\infty} \sqrt[n]{\frac{3^n}{n^3}} = \lim_{n\to\infty} \frac{3}{n^{3/n}} = \frac{3}{1} = 3.$$

Therefore by the Root Test $\sum\limits_{n=1}^{\infty} \frac{3^n}{n^3}$ diverges.

$\left(\lim\limits_{n\to\infty} n^{3/n} = 1 \text{ using the methods of} \right.$
Section 3.9: Let $y = \lim\limits_{n\to\infty} n^{3/n}$.
Then $\ln y = \lim\limits_{n\to\infty} \ln n^{3/n} = \lim\limits_{n\to\infty} \frac{3\ln n}{n}$
$\left(\text{form } \frac{\infty}{\infty}\right) = \lim\limits_{n\to\infty} \frac{\frac{3}{n}}{1} = 0.$
Thus $y = e^0 = 1.\Big)$

c) $\sum\limits_{n=1}^{\infty} \frac{(-1)^n}{n}$

$$\lim_{n\to\infty} \sqrt[n]{\left|\frac{(-1)^n}{n}\right|} = \lim_{n\to\infty} \frac{1}{\sqrt[n]{n}} = 1.$$

The Root Test fails, but we know this series converges because it is the alternating harmonic series.

Strategy for Testing Series

Concepts to Master

Applying all the previous series tests

Summary and Focus Questions

Different series lend themselves to different series tests. Follow the eight step strategy outlined in the text. Here is a brief summary of each test together with a representative example:

Test Name	**Condition(s)**	**Conclusion(s)**	**Sample**
Divergence	$\lim\limits_{n\to\infty} a_n \neq 0$	$\sum a_n$ diverges	$\sum\limits_{n=1}^{\infty} \frac{n}{n+1}$
Integral	Find $f(x)$, decreasing with $f(n) = a_n \geq 0$.	$\int_1^\infty f(x)dx$ converges if and only if $\sum a_n$ converges.	$\sum\limits_{n=1}^{\infty} \frac{\ln n}{n}$ $\int \frac{\ln x}{x}\, dx = \frac{(\ln x)^2}{2}$
p-series	$a_n = \frac{1}{n^p}$	$\sum \frac{1}{n^p}$ converges if and only if $p > 1$.	$\sum\limits_{n=1}^{\infty} \frac{1}{n^3}$
Geometric Series	$a_n = ar^{n-1}$	$\sum ar^{n-1}$ converges if and only if $|r| < 1$.	$\sum\limits_{n=1}^{\infty} \frac{2}{3^n}$
Alternating Series	$a_n = (-1)^n b_n$ $(b_n \geq 0)$	$\sum a_n$ converges if $\lim\limits_{n\to\infty} b_n = 0$ and b_n is decreasing.	$\sum\limits_{n=1}^{\infty} \frac{(-1)^n}{n}$

272

Comparison $\quad 0 \le a_n \le b_n$ $\qquad$ If $\sum b_n$ converges $\qquad$ $\displaystyle\sum_{n=1}^{\infty} \frac{1}{n^3+1}$

then $\sum a_n$ converges.

If $\sum a_n$ diverges then $\qquad$ Choose $b_n = \frac{1}{n^3}$.

$\sum b_n$ diverges.

Limit $\qquad$ $\displaystyle\lim_{n\to\infty} \frac{a_n}{b_n} = c$ $\qquad$ If $0 < c < \infty$, $\sum a_n$ $\qquad$ $\displaystyle\sum_{n=1}^{\infty} \frac{1}{2n^3+1}$

Comparison $\qquad$ converges if and only

if $\sum b_n$ does. $\qquad$ Choose $b_n = \frac{1}{n^3}$.

Ratio $\qquad$ $\displaystyle\lim_{n\to\infty} \left| \frac{a_{n+1}}{a_n} \right| = L$ $\qquad$ If $L < 1$, $\sum a_n$ $\qquad$ $\displaystyle\sum_{n=1}^{\infty} \frac{n^2}{3^n}$

converges absolutely.

If $L > 1$, $\sum a_n$

diverges.

Root $\qquad$ $\displaystyle\lim_{n\to\infty} \sqrt[n]{|a_n|} = L$ $\qquad$ If $L < 1$, $\sum a_n$ $\qquad$ $\displaystyle\sum_{n=1}^{\infty} \left(\frac{2n+1}{3n+4} \right)^n$

converges absolutely.

If $L > 1$, $\sum a_n$

diverges.

For each series, what is a good test to apply first for each?

1) $\displaystyle\sum_{n=1}^{\infty} \frac{(n+1)^n}{4^n}$.

Root Test, because of the $(n+1)^n$ and 4^n exponentials.

2) $\displaystyle\sum_{n=1}^{\infty} \frac{5}{n^2+6n+3}$.

Comparison Test, compare to $\sum \frac{1}{n^2}$.

3) $\displaystyle\sum_{n=1}^{\infty} \frac{(-1)^n n}{n^6+1}$.

Alternating Series Test for convergence, then Comparison Test (compare to $\sum \frac{1}{n^5}$).

4) $\displaystyle\sum_{n=1}^{\infty} \frac{2^n n^3}{n!}$.

Ratio Test, because of the $n!$ term.

Power Series

Concepts to Master

Power series in $(x - c)$; Interval of convergence; Radius of convergence

Summary and Focus Questions

A <u>power series in $(x - c)$</u> is an expression of the form $\sum\limits_{n=0}^{\infty} a_n(x - c)^n$, where a_n and c are constants. For any particular x, the value of the power series is an infinite series.

For some values of x it will converge while it may diverge for other values. The domain of a power series in $x - c$ is called its <u>interval of convergence</u> and contains those values of x for which the series converges. It consists of all real numbers from $c - R$ to $c + R$ for some R with $0 \leq R \leq \infty$. The number c is called the <u>center</u> and R is called the <u>radius</u> of convergence. When $R = 0$, the interval is the single point $\{c\}$. When $R = \infty$, the interval is $(-\infty, \infty)$.

The radius R is determined by applying the Ratio Test to $\sum\limits_{n=0}^{\infty} a_n(x - c)^n$.

When $0 < R < \infty$, the Ratio Test gives no information about the two endpoints $c - R$ and $c + R$. These endpoints may or may not be in the interval of convergence and must be tested separately using some other test. You should remember that the power series diverges for all x not in the interval of convergence.

1) True or False:
   A power series is an infinite series.

   False. A power series is an expression for a function, $f(x)$. For any x in the domain of f, $f(x)$ is a convergent series.

2) For what value of x does $\sum\limits_{n=0}^{\infty} a_n(x-c)^n$ always converge?

At $x = c$, $\sum\limits_{n=0}^{\infty} a_n(x-c)^n = a_0$. The power series always converges when x is the center.

3) Compute the value of:

a) $\sum\limits_{n=0}^{\infty} n^2(x-3)^n$ at $x = 4$.

When $x = 4$ the power series is $\sum\limits_{n=0}^{\infty} n^2 1^n = \sum\limits_{n=0}^{\infty} n^2$. This infinite series diverges so there is no value for $x = 4$.

b) $\sum\limits_{n=0}^{\infty} 2^n(x-1)^n$ at $x = \frac{5}{6}$.

When $x = \frac{5}{6}$ the power series is

$\sum\limits_{n=0}^{\infty} 2^n\left(-\frac{1}{6}\right)^n = \sum\limits_{n=0}^{\infty}\left(-\frac{1}{3}\right)^n$

$= 1 - \frac{1}{3} + \frac{1}{9} - \frac{1}{27} + \ldots$ This is a geometric series with $a = 1$, $r = -\frac{1}{3}$. It converges to $\dfrac{1}{1-\left(-\frac{1}{3}\right)} = \frac{3}{4}$.

4) Find the center, radius, and interval of convergence for:

a) $\sum\limits_{n=1}^{\infty} (2n)!(x-1)^n$.

Use the Ratio Test:

$\lim\limits_{n\to\infty}\left|\dfrac{(2(n+1))!(x-1)^{n+1}}{(2n)!(x-1)^n}\right|$

$= \lim\limits_{n\to\infty}(2n+1)(2n+2)|x-1| = \infty$,

except when $x = 1$. Thus the center is 1, radius is 0, and interval of convergence is $\{1\}$.

b) $\sum\limits_{n=1}^{\infty} \dfrac{(x-4)^n}{2^n n}$.

Use the Ratio Test:

$\lim\limits_{n\to\infty}\left|\dfrac{(x-4)^{n+1}}{2^{n+1}(n+1)} \cdot \dfrac{2^n n}{(x-4)^n}\right|$

$= \lim\limits_{n\to\infty}\dfrac{n}{2(n+1)}|x-4| = \frac{1}{2}|x-4|$.

From $\frac{1}{2}|x-4| < 1$ we conclude $|x-4| < 2$. The center is 4 and radius is 2. When $|x-4| = 2$, $x = 2$ or $x = 6$. At $x = 2$, the power series becomes $\sum\limits_{n=1}^{\infty}\dfrac{(-1)^n}{n}$ which converges by the Alternating Series Test. At $x = 6$ the power series is the divergent harmonic series $\sum\limits_{n=1}^{\infty}\dfrac{1}{n}$. The interval of convergence is $[2, 6)$.

c) $\displaystyle\sum_{n=1}^{\infty} \frac{x^n}{n!}$

Use the Ratio Test:

$$\lim_{n\to\infty} \left| \frac{\frac{x^{n+1}}{(n+1)!}}{\frac{x^n}{n!}} \right| = \lim_{n\to\infty} \frac{|x|}{n+1} = 0 \text{ for all values}$$

of x. Thus the interval of convergence is all real numbers, $(-\infty, \infty)$.

5) Suppose $\displaystyle\sum_{n=0}^{\infty} a_n(x-5)^n$ converges for $x = 8$. For what other values must it converge?

The interval of convergence is $5 - R$ to $5 + R$.

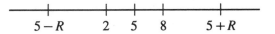

Because 8 is in this interval $8 \leq 5 + R$. Thus $3 \leq R$ which means the interval at least contains all numbers between $5 - 3$ and $5 + 3$. Thus the power series converges for at least all $x \in (2, 8]$.

Representation of Functions as Power Series

Concepts to Master

A. Representing a function by manipulating known series
B. Differentiation and integration of power series

Summary and Focus Questions

A. We will take a known function and write it as a power series. Start with a power series expression for a similar function and by substitution (such as x^2 for x) and multiplication by constants and powers of x turn the known power series into one for the given function.

Here is an example - find a power series for $\frac{2x}{1+x^2}$:

From $\frac{1}{1-x} = \sum_{n=0}^{\infty} x^n$ substitute $-x^2$ for x to obtain

$\frac{1}{1+x^2} = \sum_{n=0}^{\infty} (-x^2)^n = \sum_{n=0}^{\infty} (-1)^n x^{2n}$ and then multiply by $2x$ to obtain

$\frac{2x}{1+x^2} = \sum_{n=0}^{\infty} (-1)^n x^{2n}(2x) = \sum_{n=0}^{\infty} (-1)^n 2x^{2n+1}$. The interval of convergence

for $\frac{1}{1-x}$ is $(-1, 1)$, so from $|x| < 1$, $|x^2| < 1$ still yields $|x| < 1$.

1) Given $\frac{1}{1-x} = \sum_{n=0}^{\infty} x^n$, $|x| < 1$, find a
power series expression for:

a) $\frac{1}{2-x}$.

$\frac{1}{2-x} = \frac{1}{2\left(1-\frac{x}{2}\right)} = \frac{1}{2} \cdot \frac{1}{1-\frac{x}{2}}$. Thus

$\frac{1}{2-x} = \frac{1}{2}\sum_{n=0}^{\infty} \left(\frac{x}{2}\right)^n = \sum_{n=0}^{\infty} \frac{1}{2} \cdot \frac{x^n}{2^n}$

$= \sum_{n=0}^{\infty} \frac{x^n}{2^{n+1}}$.

From $\left|\frac{x}{2}\right| < 1$, $|x| < 2$, so $(-2, 2)$ is the interval of convergence.

b) $\frac{x}{x^3-1}$.

$\frac{x}{x^3-1} = -\frac{x}{1-x^3}$.

First substitute x^3 for x:

$$\frac{1}{1-x^3} = \sum_{n=0}^{\infty} (x^3)^n - \sum_{n=0}^{\infty} x^{3n}.$$

Now multiply by $-x$:

$$\frac{-x}{1-x^3} = -x\sum_{n=0}^{\infty} x^{3n} = -\sum_{n=0}^{\infty} x^{3n+1}.$$

Originally $|x| < 1$, hence $|x^3| < 1$. Thus the interval of convergence is still $(-1, 1)$.

B. If $f(x) = \sum_{n=0}^{\infty} a_n (x-c)^n$ has interval of convergence with radius R then:

1. f is continuous on $(c - R, c + R)$.

2. For all $x \in (c - R, c + R)$, f may be differentiated term by term:
$$f'(x) = \sum_{n=1}^{\infty} na_n (x-c)^{n-1}.$$

3. For all $x \in (c - R, c + R)$, f may be integrated term by term:
$$\int f(x)dx = C + \sum_{n=0}^{\infty} \frac{a_n(x-c)^{n+1}}{n+1}.$$

Both $f'(x)$ and $\int f(x)dx$ have radius of convergence R but the endpoints $c - R$ and $c + R$ must still be checked individually.

Term by term integration and differentiation may be used to find more power series for known functions.

For example, to find a power series for $\ln(1 + x^2)$ recall from part A that the series for $\frac{1}{1-x}$ is used to find that

$$\frac{2x}{1+x^2} = \sum_{n=0}^{\infty} (-1)^n 2x^{2n+1}.$$

Thus $\int \frac{2x}{1+x^2}\, dx = \int \sum_{n=0}^{\infty} (-1)^n 2x^{2n+1}\, dx = \sum_{n=0}^{\infty} (-1)^n 2\int x^{2n+1}\, dx$

or $\ln(1 + x^2) = \sum_{n=0}^{\infty} (-1)^n 2\frac{x^{2n+2}}{2n+2} = \sum_{n=0}^{\infty} (-1)^n \frac{x^{2n+2}}{n+1}.$

We still have $|x| < 1$ but need to check $x = 1$ and $x = -1$. At both $x = 1$ and $x = -1$, $x^{2n+2} = 1$. $\sum\limits_{n=0}^{\infty} \frac{(-1)^n}{2n+1}$ converges so the interval of convergence is $[-1, 1]$.

2) Find $f'(x)$ for
$$f(x) = \sum\limits_{n=0}^{\infty} 2^n (x - 1)^n.$$

$$f'(x) = \sum\limits_{n=1}^{\infty} 2^n n (x - 1)^{n-1}.$$

3) Find $\int f(x) dx$ where
$$f(x) = \sum\limits_{n=0}^{\infty} \frac{(x-3)^n}{n!}.$$

$$\int f(x) dx = C + \sum\limits_{n=0}^{\infty} \frac{1}{n!} \frac{(x-3)^{n+1}}{n+1}$$
$$= C + \sum\limits_{n=0}^{\infty} \frac{(x-3)^{n+1}}{(n+1)!}.$$

4) Is $f(x) = \sum\limits_{n=0}^{\infty} \frac{x^n}{4^n}$ continuous at $x = 2$?

Yes. 2 is in $[-4, 4]$, the interval of convergence for $f(x)$.

5) Given $\frac{1}{1-x} = \sum\limits_{n=0}^{\infty} x^n$, $|x| < 1$, find a power series for $\frac{1}{(1+x)^2}$.

Substitute $-x$ for x:
$$\frac{1}{1+x} = \sum\limits_{n=0}^{\infty} (-x)^n = \sum\limits_{n=0}^{\infty} (-1)^n x^n.$$
Differentiate term by term:
$$\frac{-1}{(1+x)^2} = \sum\limits_{n=0}^{\infty} (-1)^n n x^{n-1}.$$
Thus $\frac{1}{(1+x)^2} = (-1) \sum\limits_{n=0}^{\infty} (-1)^n n x^{n-1}$
$$= \sum\limits_{n=0}^{\infty} (-1)^{n+1} n x^{n-1}.$$
$|-x| < 1$ is the same as $|x| < 1$, the radius of convergence is $R = 1$. At $x = 1$ we have $\sum\limits_{n=0}^{\infty} (-1)^{n+1} n$ and at $x = -1$ we have $\sum\limits_{n=0}^{\infty} (-1)^{n+1} n (-1)^{n-1} = \sum\limits_{n=0}^{\infty} n$. Both series diverge so the interval of convergence is $(-1, 1)$.

6) Find $\int \frac{1}{1+x^5}\, dx$ as a power series.

From $\frac{1}{1-x} = \sum\limits_{n=0}^{\infty} x^n$,

$\frac{1}{1+x} = \sum\limits_{n=0}^{\infty} (-x)^n = \sum\limits_{n=0}^{\infty} (-1)^n x^n$.

Thus $\frac{1}{1+x^5} = \sum\limits_{n=0}^{\infty} (-1)^n x^{5n}$.

$\int \frac{1}{1+x^5}\, dx = \sum\limits_{n=0}^{\infty} \int (-1)^n x^{5n}\, dx$

$= \sum\limits_{n=0}^{\infty} \frac{(-1)^n}{5n+1} x^{5n+1} + C.$

Taylor and Maclaurin Series

Concepts to Master

A. Taylor series; Maclaurin series; Analytic functions
B. Taylor polynomials of degree n
C. Taylor's Formula with Remainder (Lagrange's form)
D. Multiplication and division of power series

Summary and Focus Questions

A. If a function $y = f(x)$ can be expressed as a power series it will be in this form, called the <u>Taylor series of f at c</u>:

$$f(x) = \sum_{n=0}^{\infty} \frac{f^{(n)}(C)}{n!}(x - c)^n$$

$$= f(c) + f'(c)(x - c) + \frac{f''(c)}{2!}(x - c)^2 + \ldots + \frac{f^{(n)}(c)}{n!}(x - c)^n + \ldots$$

In the special case of $c = 0$, this is called the <u>Maclaurin series of f</u>.

Not all functions can be represented by their power series. If f can be represented as a power series about c we say f is <u>analytic at c</u>. In such a case f is equal to its Taylor series at c.

Here are some basic Maclaurin series:

$$e^x = \sum_{n=0}^{\infty} \frac{x^n}{n!} \text{ for all } x. \qquad\qquad \ln(1 + x) = \sum_{n=0}^{\infty} \frac{(-1)^{n+1}}{n} x^n \text{ for } x \in (-1, 1].$$

$$\sin x = \sum_{n=0}^{\infty} \frac{(-1)^n}{(2n+1)!} x^{2n+1} \text{ for all } x. \qquad \cos x = \sum_{n=0}^{\infty} \frac{(-1)^n}{(2n)!} x^{2n} \text{ for all } x.$$

$$\frac{1}{1-x} = \sum_{n=0}^{\infty} x^n \text{ for } |x| < 1. \qquad\qquad \tan^{-1} x = \sum_{n=0}^{\infty} (-1)^n \frac{x^{2n+1}}{2n+1} \text{ for } |x| \le 1.$$

281

To find a Taylor series for a given $y = f(x)$, you must find a formula for $f^{(n)}(c)$ usually in terms of n. Computing the first few derivatives $f'(c)$, $f''(c)$, $f'''(c)$, $f^{(4)}(c), \ldots$ often helps.

For some functions f it is easier to find the Taylor series for $f'(x)$ or $\int f(x)dx$ then integrate or differentiate term by term to obtain the series for f. In other cases, substitutions in the basic Maclaurin series can be used to find the Taylor series for f.

1) Find the Taylor series for $\frac{1}{x}$ at $c = 1$.

   a) directly from the definition.

We must find the general form of $f^{(n)}(1)$:
$$f(x) = x^{-1} \qquad f(1) = 1$$
$$f'(x) = -x^{-2} \qquad f'(1) = -1$$
$$f''(x) = 2x^{-3} \qquad f''(1) = 2$$
$$f'''(x) = -6x^{-4} \qquad f'''(1) = -6$$
$$f^{(4)}(x) = 24x^{-5} \qquad f^{(4)}(1) = 24$$
In general, $f^{(n)}(1) = (-1)^n n!$
Thus the Taylor series is
$$\sum_{n=0}^{\infty} \frac{(-1)^n n!}{n!}(x-1)^n = \sum_{n=0}^{\infty} (-1)^n (x-1)^n$$
$$= \sum_{n=0}^{\infty} (1-x)^n.$$

   b) using substitution in a geometric series $\left(\text{Hint: } \frac{1}{x} = \frac{1}{1-(1-x)}\right)$.

In the form $\frac{1}{1-(1-x)}$, this is the value of a geometric series with $a = 1$, $r = 1 - x$.
Thus $\frac{1}{x} = \sum_{n=1}^{\infty} 1(1-x)^{n-1} = \sum_{n=0}^{\infty} (1-x)^n.$

   c) What is the interval of convergence for your answer to part a)?

$$\lim_{n\to\infty} \left| \frac{(1-x)^{n+1}}{(1-x)^n} \right| = \lim_{n\to\infty} |1-x| = |1-x|.$$
$|1-x| < 1$ is equivalent to $0 < x < 2$.
At $x = 0$ and $x = 2$ the terms of
$$\sum_{n=0}^{\infty} (1-x)^n$$
do not approach zero so the series diverges. The interval of convergence is $(0, 2)$.

2) Find directly the Maclaurin series for $f(x) = e^{4x}$.

$$f(x) = e^{4x} \qquad f(0) = 1$$
$$f'(x) = 4e^{4x} \qquad f'(0) = 4$$
$$f''(x) = 16e^{4x} \qquad f''(0) = 16$$

In general $f^{(n)}(0) = 4^n$.

Thus $e^{4x} = \sum\limits_{n=0}^{\infty} \frac{4^n}{n!} x^n$.

3) True, False:
If f is analytic at c, then
$$f(x) = \sum_{n=0}^{\infty} \frac{f^{(n)}(c)}{n!} (x - c)^n.$$

True.

4) Obtain the Maclaurin series for $\frac{x}{1+x^2}$ from the series for $\frac{1}{1-x} = \sum\limits_{n=0}^{\infty} x^n$.

$\frac{1}{1-x} = 1 + x + x^2 + \ldots + x^n + \ldots$
Substitute $-x^2$ for x:
$\frac{1}{1+x^2} = 1 - x^2 + x^4 - x^6 + \ldots$
$\qquad + (-1)^n x^{2n} + \ldots$
Now multiply by x:
$\frac{x}{1+x^2} = x - x^3 + x^5 - x^7 + \ldots$
$\qquad + (-1)^n x^{2n+1} + \ldots$
$\qquad = \sum\limits_{n=0}^{\infty} (-1)^n x^{2n+1}$

5) Using $e^x = \sum\limits_{n=0}^{\infty} \frac{x^n}{n!}$ find the Maclaurin series for $\sinh x$.
(Remember, $\sinh x = \frac{e^x - e^{-x}}{2}$.)

$e^x = 1 + x + \frac{x^2}{2!} + \ldots + \frac{x^n}{n!} + \ldots$
Thus $e^{-x} = 1 - x + \frac{x^2}{2!} + \ldots$
$\qquad + \frac{(-1)^n x^n}{n!} + \ldots$
and subtracting term by term (the even terms cancel)
$e^x - e^{-x} = 2x + \frac{2x^3}{3!} + \ldots + \frac{2x^{2n+1}}{(2n+1)!} + \ldots$
Thus $\sinh x = x + \frac{x^3}{3!} + \ldots$
$\qquad + \frac{x^{2n+1}}{(2n+1)!} + \ldots = \sum\limits_{n=0}^{\infty} \frac{x^{2n+1}}{(2n+1)!}.$

6) Find the infinite series expression for $\int_0^{1/2} \frac{1}{1-x^3}\, dx$.

$\frac{1}{1-x} = \sum\limits_{n=0}^{\infty} x^n$. Thus $\frac{1}{1-x^3} = \sum\limits_{n=0}^{\infty} x^{3n}$.

$\int \frac{1}{1-x^3}\, dx = \int \sum\limits_{n=0}^{\infty} x^{3n}\, dx = \sum\limits_{n=0}^{\infty} \frac{x^{3n+1}}{3n+1} F(x).$
$\int_0^{1/2} \frac{1}{1-x^3}\, dx = F\left(\frac{1}{2}\right) - F(0)$

$\qquad = \sum\limits_{n=0}^{\infty} \frac{\left(\frac{1}{2}\right)^{3n+1}}{3n+1} - 0 = \sum\limits_{n=0}^{\infty} \frac{1}{(6n+2)8^n}.$

B. If $f^{(n)}(c)$ exists, the <u>Taylor polynomial of degree n for f about c</u> is

$$T_n(x) = f(c) + \frac{f'(c)}{1!}(x-c) + \frac{f''(c)}{2!}(x-c)^2 + \ldots \frac{f^{(n)}(c)}{n!}(x-c)^n.$$

$T_n(x)$ is simply the nth partial sum of the Taylor series for $f(x)$.

$T_1(x) = f(c) + f'(c)(x - c)$ is the familiar equation for the tangent line to $y = f(x)$ at $x = c$. $T_2(x)$ the "tangent parabola."

Since the first n derivatives of T_n at c are equal to the corresponding first n derivatives of f at c, $T_n(x)$ is an approximation for $f(x)$ if x is near c. Recall that we used $T_1(x)$ and $T_2(x)$ for approximations in Chapter 2.

7) Let $f(x) = \sqrt{x}$.

    a) Construct the Taylor polynomial of degree 3 for $f(x)$ about $x = 1$.

$$f(x) = x^{1/2} \qquad f(1) = 1$$
$$f'(x) = \tfrac{1}{2}x^{-1/2} \qquad f'(1) = \tfrac{1}{2}$$
$$f''(x) = -\tfrac{1}{4}x^{-3/2} \qquad f''(1) = -\tfrac{1}{4}$$
$$f^{(3)}(x) = \tfrac{3}{8}x^{-5/2} \qquad f^{(3)}(1) = \tfrac{3}{8}$$
$$T_3(x)$$
$$= 1 + \tfrac{1}{2}(x-1) - \tfrac{1/4}{2!}(x-1)^2 + \tfrac{3/8}{3!}(x-1)^3$$
$$= 1 + \tfrac{1}{2}(x-1) - \tfrac{1}{8}(x-1)^2 + \tfrac{1}{16}(x-1)^3$$

    b) Approximate $\sqrt{\tfrac{3}{2}}$ using the Taylor polynomial of degree 3 from part a).

$$T_3\left(\tfrac{3}{2}\right) = 1 + \tfrac{1}{2}\left(\tfrac{1}{2}\right) - \tfrac{1}{8}\left(\tfrac{1}{2}\right)^2 + \tfrac{1}{16}\left(\tfrac{1}{2}\right)^3$$
$$= \tfrac{157}{128} \approx 1.2265.$$
$$\left(\text{To 4 decimals } \sqrt{\tfrac{3}{2}} = 1.2247.\right)$$

8) a) Find the Taylor polynomial of degree 6 for $f(x) = \cos x$ about $x = 0$.

$$f(0) = \cos 0 = 1$$
$$f'(x) = -\sin x, \ f'(0) = 0$$
$$f''(x) = -\cos x, \ f''(0) = -1$$
$$f^{(3)}(x) = \sin x, \ f^{(3)}(0) = 0$$
$$f^{(4)}(x) = \cos x, \ f^{(4)}(0) = 1$$
$$f^{(5)}(0) = 0 \text{ and } f^{(6)}(0) = -1$$
Thus $T_6(x) = 1 + \tfrac{-1}{2!}x^2 + \tfrac{1}{4!}x^4 - \tfrac{1}{6!}x^6$
$$T_6(x) = 1 - \tfrac{x^2}{2} + \tfrac{x^4}{24} - \tfrac{x^6}{720}.$$
Note, of course, that this is the first few terms of the Maclaurin series for cosine.

b) Use your answer to part a) to
approximate cos 1.

$$\cos 1 \approx T_6(1) = 1 - \frac{(1)^2}{2} + \frac{(1)^4}{24} - \frac{(1)^6}{720}$$
$$= 1 - 0.5 + 0.041667 - 0.001389$$
$$= 0.540278$$

(Note: $\cos 1 = 0.540302$ to 6 decimal places.)

C. Taylor's Formula with Remainder (Lagrange's form):

Let f be a function whose $n + 1$ derivatives exist and are continuous in an interval I containing c. Then for $x \in I$,
$$f(x) = T_n(x) + R_n(x),$$
$$R_n(x) = \frac{f^{(n+1)}(z)}{(n+1)!}(x - c)^{n+1} \text{ for some number } z \text{ between } c \text{ and } x.$$

$R_n(x)$ is expressed in <u>Lagrange's form</u> for the remainder and is the <u>error</u> in approximating $f(x)$ with $T_n(x)$.

If f has derivatives of all orders and $\lim\limits_{n \to \infty} R_n(x) = 0$, then $f(x)$ is equal to its Taylor series on its interval of convergence. Thus the approximation $T_n(x)$ often can be made accurate to within any given number ε by selecting n such that $|R_n(x)| < \varepsilon$. This usually requires finding upper bound estimates for $\left| f^{(n+1)}(z) \right|$.

9) True or False:
In general, the larger the number n, the closer the Taylor polynomial approximation is to the actual functional value.

True.

10) a) Write Taylor's Formula with
Remainder with $n = 3$ for
$f(x) = x^{5/2}$ about $c = 4$.

$$f(x) = x^{5/2} \qquad\qquad f(4) = 32$$
$$f'(x) = \tfrac{5}{2}x^{3/2} \qquad\quad f'(4) = 20$$
$$f''(x) = \tfrac{15}{4}x^{1/2} \qquad\; f''(4) = \tfrac{15}{2}$$
$$f'''(x) = \tfrac{15}{8}x^{-1/2} \qquad f'''(4) = \tfrac{15}{16}$$

$$T_3(x) = 32 + 20(x-4) + \frac{15/2}{2!}(x-4)^2$$
$$+ \frac{15/16}{3!}(x-4)^3$$
$$= 32 + 20(x-4) + \frac{15}{4}(x-4)^2$$
$$+ \frac{5}{32}(x-4)^3.$$

Since $f^{(4)}(x) = -\frac{15}{16}x^{-3/2}$,

$R_3(x) = \frac{-(15/16)z^{-3/2}}{4!}(x-4)^4$. Therefore

$R_3(x) = \frac{-5}{128z^{3/2}}(x-4)^4$, for z between c and x. Taylor's Formula is

$f(x) = T_3(x) + R_3(x)$, where $T_3(x)$ and $R_3(x)$ are given above.

b) Estimate the error between $f(5) = 5^{5/2}$ and $T_3(5)$.

An estimate of the error is

$|R_3(5)| = \frac{5}{128z^{3/2}}(5-4)^4 = \frac{5}{128z^{3/2}}$.

Since $4 < z < 5$, $8 = 4^{3/2} < z^{3/2}$.
Thus $|R_3(5)| < \frac{5}{128(8)} = \frac{5}{1024} \approx 0.00488$.

c) Compute $T_3(5)$.

$T_3(5) = 32 + 20(1)^1 + \frac{15}{4}(1)^2 + \frac{5}{32}(1)^3$
$= \frac{1789}{32} \approx 55.9063$.

d) To four decimal places, what is the actual error?

To four decimals $f(5) = 5^{5/2} = 55.9017$.
Thus the error is
$55.9063 - 55.9017 = 0.0046$, which is within 0.00488.

11) Find $\int_0^1 xe^{-x}\, dx$ using series to within 0.01.

$\left(\text{We could use integration by parts:}\right.$

$\int_0^1 xe^{-x}\, dx = (-xe^{-x} - e^{-x})\Big|_0^1 = 1 - \frac{2}{e}.\Big)$

$e^x = \sum_{n=0}^{\infty} \frac{x^n}{n!}$, $e^{-x} = \sum_{n=0}^{\infty} \frac{(-1)^n x^n}{n!}$,

so $xe^{-x} = \sum_{n=0}^{\infty} \frac{(-1)^n x^{n+1}}{n!}$. Thus

$\int_0^1 xe^{-x}\, dx = \int_0^1 \sum_{n=0}^{\infty} \frac{(-1)^n x^{n+1}}{n!}\, dx$

$= \sum_{n=0}^{\infty} \frac{(-1)^n x^{n+2}}{n!(n+2)}\Big|_0^1 = \sum_{n=0}^{\infty} \frac{(-1)^n}{n!(n+2)} - 0$

$= \sum_{n=0}^{\infty} \frac{(-1)^n}{n!(n+2)}$

$= \frac{1}{0!2} - \frac{1}{1!3} + \frac{1}{2!4} - \frac{1}{3!5} + \frac{1}{4!6}$
$- \frac{1}{5!7} + \cdots$

$= \frac{1}{2} - \frac{1}{3} + \frac{1}{8} - \frac{1}{30} + \frac{1}{144}$.

This is an alternating series and the fifth term $\frac{1}{144}$ is less than 0.01. Thus we may use $\int_0^1 xe^{-x}\,dx \approx \frac{1}{2} - \frac{1}{3} + \frac{1}{8} - \frac{1}{30} \approx 0.2583$. This is within 0.01 of the actual value $1 - \frac{2}{e} \approx 0.2642$. We note in passing that the series $\sum\limits_{n=0}^{\infty} \frac{(-1)^n}{n!(n+2)} = 1 - \frac{2}{e}$.

D. Convergent series may be multiplied and divided like polynomials. Often finding the first few terms of the result is sufficient.

12) Find the first three terms of the Maclaurin series for $e^x \sin x$.

$$e^x = 1 + x + \frac{x^2}{2} + \frac{x^3}{6} + \dots$$
$$\sin x = x - \frac{x^3}{3!} + \frac{x^5}{5!} - \dots$$

Thus

$$e^x \sin x = x\left(1 + x + \frac{x^2}{2} + \frac{x^3}{6} + \dots\right)$$
$$- \frac{x^3}{6}\left(1 + x + \frac{x^2}{2} + \frac{x^3}{6} + \dots\right)$$
$$+ \frac{x^5}{120}\left(1 + x + \frac{x^2}{2} + \frac{x^3}{6} + \dots\right)$$
$$+ \dots$$
$$= \left(x + x^2 + \frac{x^3}{2} + \frac{x^4}{6} + \dots\right)$$
$$- \left(\frac{x^3}{6} + \frac{x^4}{6} + \frac{x^5}{12} + \frac{x^6}{36} + \dots\right)$$
$$+ \left(\frac{x^5}{120} + \frac{x^6}{120} + \frac{x^7}{240} + \dots\right)$$
$$= x + x^2 + \frac{x^3}{6} + \dots$$

Applications of Taylor Polynomials

Concepts to Master

Approximating function values with Taylor polynomials; Estimating error in approximations

Summary and Focus Questions

The nth degree Taylor polynomial for a function $f(x)$ at c,
$$T_n(x) = f(c) + f'(c)(x-c) + \frac{f''(c)}{2!}(x-c)^2 + \ldots + \frac{f^{(n)}(c)}{n!}(x-c)^n,$$
may be used to approximate $f(x)$ for any given x in the radius of convergence of the Taylor series for f. How good the approximation $f(x) \approx T_n(x)$ will be dependent on:
   1. n - the larger n is, the better the estimate, and
   2. x and c - the closer x is to c, the better the estimate.

This is shown in the remainder $R_n = f(x) - T_n(x)$ by Taylor's formula:
$$R_n(x) = \frac{f^{(n+1)}(z)}{(n+1)!}(x-c)^{n+1} \text{ where } z \text{ is between } x \text{ and } c.$$

$|R_n(x)|$ may be calculated or a graphing calculator can be used to estimate it. Or, in the special case where the Taylor series is an Alternating Series, $|R_n|$ may be estimated with $\left|\frac{f^{n+1}(x)}{(n+1)!}(x-c)^{n+1}\right|$, the $(n+1)$st term of the Taylor series.

1)   a)   For $f(x) = \frac{1}{3+x}$, find $T_3(x)$ where $c = 2$.

$f(x) = (3+x)^{-1}$ $f(2) = \frac{1}{5}$

$f'(x) = -(3+x)^{-2}$ $f'(2) = \frac{-1}{25}$

$f''(x) = 2(3+x)^{-3}$ $f''(2) = \frac{2}{125}$

$f'''(x) = -6(3+x)^{-4}$ $f'''(2) = \frac{-6}{625}$

$T_3(x) = \frac{1}{5} - \frac{1}{25}(x-2) + \frac{2/125}{2!}(x-2)^2$

$\qquad - \frac{6/625}{3!}(x-2)^3$

$\qquad = \frac{1}{5} - \frac{1}{25}(x-2) + \frac{1}{125}(x-2)^3$

$\qquad - \frac{1}{625}(x-2)^3$

b) Use Taylor's formula to estimate the accuracy of $f(x) \approx T_3(x)$ when $1 \leq x \leq 4$.

From part a), $f^{(4)}(x) = 24(3+x)^{-5}$.
$$|f(x) - T_3(x)| = |R_3(x)|$$
$$= \left| \frac{24(3+z)^{-5}}{4!} (x-2)^4 \right| = \frac{(x-2)^4}{(3+z)^5}$$
where z is between x and 2.
From $1 \leq x \leq 4$, $-1 \leq x - 2 \leq 2$.
Thus $(x-2)^4 \leq 2^4 = 16$.
z is between 2 and x, so $1 \leq z \leq 4$,
$4 \leq 3 + z \leq 7$. Thus $4^5 \leq (3+z)^5 \leq 7^5$,
or $\frac{1}{4^5} \geq \frac{1}{(3+z)^5} \geq \frac{1}{7^5}$. Therefore
$$|R_3(x)| = \frac{(x-2)^4}{(3+z)^5} \leq \frac{16}{4^5} = \frac{1}{64} \approx 0.0156.$$

c) Check the accuracy of the estimate when $x = 3$.

$$f(3) = \frac{1}{3+3} = \frac{1}{6} \approx 0.1667$$
$$T_3(3) = \frac{1}{5} - \frac{1}{25} + \frac{1}{125} - \frac{1}{625} = 0.1664$$
Indeed $|f(3) - T_3(3)| < 0.0156$.

2) a) What degree Maclaurin polynomial is needed to approximate $\cos 0.5$ accurate to within 0.00001?

Rephrased, the question asks: For what n is $|R_n(0.5)| < 0.00001$ when $f(x) = \cos x$ and $c = 0$?
$$|R_n(0.5)| = \left| \frac{f^{(n+1)}(z)}{(n+1)!} (0.5 - 0)^{n+1} \right|$$
$$= \frac{|f^{(n+1)}(z)|(0.5)^{n+1}}{(n+1)!} \leq \frac{(0.5)^{n+1}}{(n+1)!}$$
since $\left| f^{(n+1)}(z) \right| \leq 1$.
Now $\frac{(0.5)^{n+1}}{(n+1)!} > 0.00001$ for $n = 1, 2, 3, 4$.
However, for $n = 5$,
$$\frac{(0.5)^{n+1}}{(n+1)!} = 0.000022 < 0.00001.$$
Thus the degree should be at least 5.

b) Use your answer to part a) to estimate $\cos 0.5$ to within 0.00001.

For $f(x) = \cos x$, $f(0) = 1$,
$f'(0) = 0$, $f''(0) = -1$,
$f^{(3)}(0) = 0$, $f^{(4)}(0) = 1$, and
$f^{(5)}(0) = 0$.
Thus $T_5(x) = 1 - \frac{x}{2!} + \frac{x}{4!}$ and
$T_5(0.5) \approx 0.87760$.

On Your Own

Review and Preview 1

_____ 1. True, False: $x^2 + 6x + 2y = 1$ defines y as a function of x.

_____ 2. True, False: In $V = \frac{4}{3}\pi r^3$, r is the dependent variable.

_____ 3. The implied domain of $f(x) = \frac{1}{\sqrt{1-x}}$ is:

     a) $(1, \infty)$ c) $x \neq 1$

     b) $(-\infty, 1)$ d) $(-1, 1)$

_____ 4. True, False: The graph below is the graph of an even function.

_____ 5. Which graph best represents $y = x + \frac{1}{x}$?

a)   b)   c)

_____ 6. For $f(x) = 3x + 1$ and $g(x) = x$, $\frac{f}{g}(x) =$ _____ ?

     a) $\frac{3+x}{x}$ c) $\frac{3x+1}{x}$

     b) $\frac{x}{3x+1}$ d) $3x + \frac{1}{x}$

_____ **7.** For $f(x) = \sqrt{x^2 + x}$, we may write $f(x) = h \circ g(x)$, where:

    a) $h(x) = \sqrt{x}$ and $g(x) = x^2 + x$
    b) $h(x) = x^2 + x$ and $g(x) = \sqrt{x}$
    c) $h(x) = x^2$ and $g(x) = \sqrt{x}$
    d) $h(x) = x^2 + x$ and $g(x) = x^2$

_____ **8.** Let $f(x) = 2 + \sqrt{x}$ and $g(x) = x + 3$. Then $g \circ f(x) =$ _____ ?

    a) $2 + \sqrt{x + 3}$ c) $3 + \sqrt{x + 2}$
    b) $2 + \sqrt{x + 3}$ d) $(x + 3)(2 + \sqrt{x})$

_____ **9.** Sometimes, Always, or Never: $f \circ g(x) = g \circ f(x)$.

_____ **10.** For $f(x) = \cos(x^2 + 1)$ we may write $f(x) = h \circ g(x)$, where:

    a) $h(x) = \cos x^2$ and $g(x) = x + 1$
    b) $h(x) = \cos x$ and $g(x) = x^2 + 1$
    c) $h(x) = x^2 + 1$ and $g(x) = \cos x$
    d) $h(x) = x^2$ and $g(x) = \cos(x + 1)$

Review and Preview 4

_____ 1. The four general steps to solve problems are:

    a) 1. Plan   b) 1. Understand problem c) 1. Draw a figure
        2. Calculate   2. Plan   2. Label the figure
        3. Execute   3. Carry out   3. Calculate
        4. Review   4. Look back   4. Evaluate

_____ 2. Frick and Frack are on each end of a 10 foot long teeter-totter. Frick weighs 85 pounds and Frack weighs 110. Where should the fulcrum (the balance point) be so that the teeter-totter balances when the two boys are on it?

    a) 6.54 ft. from Frick's end   c) 5.95 ft. from Frick's end
    b) 5.64 ft. from Frick's end   d) 4.46 ft. from Frick's end

Section 1.1

_____ **1.** For $f(x) = 5x^2 + 1$ the slope of the secant line between the points corresponding to $x = 3$ and $x = 4$ is:

       a) 7   b) 35   c) 81   d) 1

_____ **2.** A guess for the slope of the tangent line to $f(x) = 5x^2 + 1$ at $P(3,\ 26)$ based on the chart below is:

x	4	3.1	3.01	3.001
$f(x)$	81	49.05	46.3005	46.030005

       a) 26   b) 30   c) 35   d) 46

_____ **3.** True, False:

The slope of a tangent line may be interpreted as average velocity.

1. In the graph at the right;
$\lim\limits_{x \to 2} f(x) =$ _____.

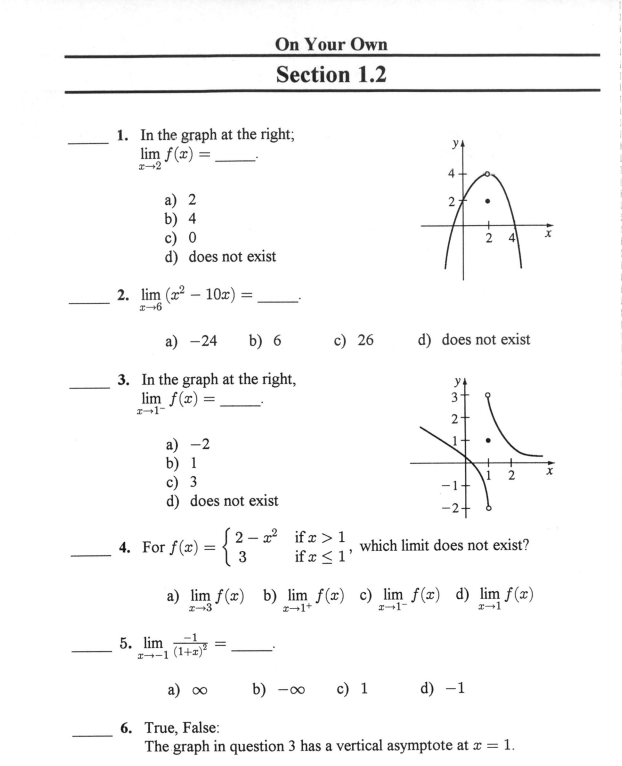

a) 2
b) 4
c) 0
d) does not exist

2. $\lim\limits_{x \to 6} (x^2 - 10x) =$ _____.

a) -24 b) 6   c) 26   d) does not exist

3. In the graph at the right,
$\lim\limits_{x \to 1^-} f(x) =$ _____.

a) -2
b) 1
c) 3
d) does not exist

4. For $f(x) = \begin{cases} 2 - x^2 & \text{if } x > 1 \\ 3 & \text{if } x \leq 1 \end{cases}$, which limit does not exist?

a) $\lim\limits_{x \to 3} f(x)$ b) $\lim\limits_{x \to 1^+} f(x)$ c) $\lim\limits_{x \to 1^-} f(x)$ d) $\lim\limits_{x \to 1} f(x)$

5. $\lim\limits_{x \to -1} \frac{-1}{(1+x)^2} =$ _____.

a) ∞ b) $-\infty$ c) 1   d) -1

6. True, False:
The graph in question 3 has a vertical asymptote at $x = 1$.

Section 1.3

_____ **1.** True, False:
If $h(x) = g(x)$ for all $x \neq a$ and $\lim_{x \to a} h(x) = L$, then $\lim_{x \to a} g(x) = L$.

_____ **2.** $\lim_{x \to 3} \frac{7x+6}{5x-12} =$ _____.

a) $-\frac{1}{2}$ b) $\frac{7}{5}$ c) 9   d) 4

_____ **3.** Sometimes, Always, or Never: $\lim_{x \to a} \frac{f(x)}{g(x)} = \frac{\lim_{x \to a} f(x)}{\lim_{x \to a} g(x)}$.

_____ **4.** $\lim_{x \to 2} \frac{x^2(x-1)}{1-x} =$ _____.

a) -4 b) -1 c) 0   d) does not exist

_____ **5.** $\lim_{x \to 1} \frac{x^3 - x^2}{1 - x} =$ _____.

a) 1   b) -1 c) 0   d) does not exist

_____ **6.** If $2x + 2 \leq f(x) \leq x^2 + 3$, then $\lim_{x \to 1} f(x) =$ _____.

a) 1
b) 4
c) does not exist
d) exists but cannot be determined from the information given

_____ 1. True, False:
$$\lim_{x \to a} \frac{f(x)-f(a)}{x-a} = \lim_{k \to 0} \frac{f(a+k)-f(a)}{k}.$$

_____ 2. True, False:
$\frac{f(x)-f(a)}{x-a}$ may be interpreted as the instantaneous velocity of a particle at time a.

_____ 3. The slope of the tangent line to $y = x^3$ at $x = 2$ is:

   a) 18   b) 12   c) 6   d) 0

_____ 4. True, False:
Slope of a tangent line, instantaneous velocity, and instantaneous rate of change of a function are each an interpretation of the same limit concept.

_____ 5. Given the table of function values below, which of the following is the best estimate for the instantaneous rate of change of $y = f(x)$ at $x = 4$?

x	6	5	4.5	4.1	4.05	4.01	4
$f(x)$	17	13.3	11.7	10.32	10.16	10.031	10

   a) 17   b) 10   c) 3   d) 0

_____ 1. True, False: $f'(x) = \lim\limits_{x \to 0} \frac{f(x+h)-f(x)}{h}$.

_____ 2. True, False: $f'(x) = \lim\limits_{x \to a} \frac{f(x)-f(a)}{x-a}$.

_____ 3. For $f(x) = 10x^2$, $f'(3) = $ _____.

     a) 10   b) 20   c) 30   d) 60

_____ 4. The instantaneous rate of change of $y = x^3 + 3x$ at the point corresponding to $x = 2$ is:

     a) 10   b) 15   c) 2   d) 60

_____ 5. $\lim\limits_{h \to 0} \frac{\sqrt{4+h}-2}{h}$ is the derivative of:

     a) $f(x) = \sqrt{4+x}$ at $x = 2$
     b) $f(x) = \sqrt{x}$ at $x = 4$
     c) $f(x) = \sqrt{x}$ at $x = 2$
     d) $f(x) = \frac{\sqrt{4+h}-2}{h}$ at $x = 0$

_____ 6. True, False: $\frac{dm}{dv}$ stands for the derivative of the function m with respect to the variable v.

Section 2.2

_____ **1.** For $f(x) = 6x^4$, $f'(x) =$ _____.

     a) $24x$ b) $6x^3$ c) $10x^3$ d) $24x^3$

_____ **2.** For $y = \frac{1}{2}gt^2$ (where g is a constant), $\frac{dy}{dt} =$ _____.

     a) $\frac{1}{4}g^2$ b) $\frac{1}{2}g$ c) gt d) gt^2

_____ **3.** True, False: $(f(x)g(x))' = f'(x)g'(x)$.

_____ **4.** For $f(x) = \frac{x}{x+1}$, $f'(x) =$ _____.

     a) 1 b) $\frac{x^2}{(x+1)^2}$ c) $\frac{-1}{(x+1)^2}$ d) $\frac{1}{(x+1)^2}$

_____ **5.** For $f(x) = 5 + g(x)$, $f'(x) =$ _____.

     a) $5g'(x)$ b) $5 + g'(x)$ c) $0 \cdot g'(x)$ d) $g'(x)$

Section 2.3

_____ **1.** True, False: $f'(t)$ is used to measure the average rate of change of f with respect to t.

_____ **2.** Air is being pumped into a chamber so that after t seconds the pressure in the chamber is $3 + 3x + x^2$ pounds/in². The rate of change of the pressure at $t = 2$ second is:

     a) 7 pounds/in²/sec
     b) $3\frac{1}{6}$ pounds/in²/sec
     c) $3\frac{1}{2}$ pounds/in²/sec
     d) $2\frac{5}{6}$ pounds/in²/sec

_____ **3.** A bacteria population grows in such a way that after t hours its population is $1000 + 10t + t^3$. The growth rate after 3 hours is:

     a) 13 bacteria/hour
     b) 37 bacteria/hour
     c) 10 bacteria/hour
     d) 1057 bacteria/hour

_____ **4.** A circle starts with radius 0 and increases in area. The rate of change of area with respect to its radius r is:

     a) $2\pi r$ b) πr^2 c) πr d) $2\pi r^2$

_____ **1.** $\lim\limits_{x \to 0} \frac{2 \sin x}{3x} = $ __ __.

a) 1   b) $\frac{2}{3}$ c) $\frac{3}{2}$ d) 0

_____ **2.** If $\lim\limits_{x \to 0} \frac{\sin x}{x} = 1$, the angle x must be measured in:

a) radians b) degrees c) it does not matter

_____ **3.** For $y = x \sin x$, $y' = $ _____.

a) $x \cos x$
b) $x \cos x + 1$
c) $\cos x$
d) $x \cos x + \sin x$

_____ **4.** For $y = 3x + \tan x$, $y' = $ _____.

a) $3 + \sec x$
b) $3x \sec^2 x + 3 \tan x$
c) $3 + \sec^2 x$
d) $x \cos x + \sin x$

_____ **5.** For $y = \sin^2 x + \cos^2 x$, $y' = $ _____.

a) $2 \sin x - 2 \cos x$
b) $2 \sin x \cos x$
c) $4 \sin x \cos x$
d) 0

Section 2.5

_____ **1.** True, False: If $y = f(g(x))$ then $y' = f'(g'(x))$.

_____ **2.** For $y = (x^2 + 1)^2$, $y' =$ _____.

       a) $2(x^2 + 1)$ b) $4(x^2 + 1)$ c) $4x(x^2 + 1)$ d) $2x(x^2 + 1)$

_____ **3.** For $y = \sin 6x$, $\frac{dy}{dx} =$ _____.

       a) $\cos 6x$ b) $6 \cos x$ c) $\cos 6$ d) $6 \cos 6x$

_____ **4.** For $y = \frac{4}{\sqrt{3+2x}}$, $y' =$ _____.

       a) $-4(3 + 2x)^{-3/2}$
       b) $4(3 + 2x)^{-3/2}$
       c) $-2(3 + 2x)^{-1/2}$
       d) $2(3 + 2x)^{-1/2}$

_____ **5.** True, False: For $y = f(u)$ and $u = g(x)$, $\frac{dy}{dx} = \frac{df}{du} \cdot \frac{du}{dx}$.

_____ 1. True, False: $y = \sqrt[3]{4 - \sqrt{x^2 + 1}}$ defines y implicitly as a function of x.

_____ 2. If y is defined as a function of x by $y^3 = x^2$, $y' = $ _____.

a) $\frac{2x}{y^3}$ b) $\frac{2x}{3y^2}$ c) $\frac{x^2}{3y^2}$ d) $\frac{x^2}{3y}$

_____ 3. The slope the line tangent to the circle $x^2 + y^2 = 100$ at the point $(-6, 8)$ is:

a) $\frac{3}{4}$ b) $-\frac{3}{4}$ c) $\frac{4}{3}$ d) $-\frac{4}{3}$

_____ **1.** True, False: $y^{(6)}$ is the derivative of $y^{(5)}$.

_____ **2.** For $y = \frac{5}{x^3}$, $y'' = $ _____.

a) $\frac{5}{6x}$ b) $\frac{5}{12x}$ c) $-\frac{60}{x^5}$ d) $60x^{-5}$

_____ **3.** For $y = x^n$ (n a positive integer), $y^{(n)} = $ _____.

a) 0 b) $n!$ c) nx^{n-1} d) $n!x$

_____ **4.** For a particle whose position at time t (seconds) is $t^3 - 3t^2 + 10t + 1$ meters, its acceleration at time $t = 3$ is:

a) 18 m/sec^2 b) 12 m/sec^2 c) 0 m/sec^2 d) -5 m/sec^2

_____ **5.** For $y = \cos x$, $y^{(14)} = $ _____.

a) $\cos x$ b) $\sin x$ c) $-\cos x$ d) $-\sin x$

_____ 1. A right triangle has one leg with constant length 8 cm. The length of the other leg is decreasing at 3 cm/sec. The rate of change of the hypotenuse when the variable leg is 6 cm is.

 a) $-\frac{9}{5}$ cm/sec
 b) 3 cm/sec
 c) $-\frac{5}{3}$ cm/sec
 d) $\frac{5}{3}$ cm/sec

_____ 2. How fast is the angle between the hands of a clock increasing at 3:30 pm?

 a) 2π radians/hr
 b) $\frac{\pi}{6}$ radians/hr
 c) $\frac{11\pi}{6}$ radians/hr
 d) $\frac{5\pi}{6}$ radians/hr

Section 2.9

_____ **1.** For $f(x) = x^3 + 7x$, $df =$ _____.

    a) $(x^3 + 7x)\,dx$
    b) $(3x^2 + 7)\,dx$
    c) $3x^2 + 7$
    d) $x^3 + 7x + dx$

_____ **2.** The best approximation of Δy using differentials for $y = x^2 - 4x$ at $x = 6$ when $\Delta x = dx = 0.03$ is:

    a) 0.36   b) 0.24   c) 0.20   d) 0.30

_____ **3.** The linearization of $f(x) = \frac{1}{\sqrt{x}}$ at $x = 4$ is:

    a) $L(x) = \frac{1}{2} + \frac{1}{\sqrt{x}}(x - 2)$

    b) $L(x) = \frac{1}{4} + \frac{1}{2}(x - 2)$

    c) $L(x) = \frac{1}{2} + \frac{1}{2}(x - 4)$

    d) $L(x) = \frac{1}{2} - \frac{1}{16}(x - 4)$

_____ **4.** Using differentials, an approximation to $(3.04)^3$ is:

    a) 28.094464   b) 28.08   c) 28.04   d) 28

_____ **5.** The The quadratic approximation for $f(x) = x^6$ at $x = 1$ is:

    a) $P(x) = 6(x - 1) + 30(x - 1)^2$
    b) $P(x) = 1 + 6(x - 1) + 30(x - 1)^2$
    c) $P(x) = 6(x - 1) + 15(x - 1)^2$
    d) $P(x) = 1 + 6(x - 1) + 15(x - 1)^2$

_____ **1.** Sometimes, Always, or Never:
The value of x_2 in Newton's Method will be a closer approximation to the root than the initial value x_1.

_____ **2.** With an initial estimate of -1, use Newton's Method once to estimate a zero of $f(x) = 5x^2 + 10x + 3$.

   a) $-\frac{6}{5}$ b) $-\frac{4}{5}$ c) $-\frac{7}{5}$ d) $-\frac{3}{5}$

_____ **3.** Suppose in an estimation of a root of $f(x) = 0$ we make an initial guess x_1, so that $f'(x_1) = 0$. Which of the following will be true?

   a) x_1 is a root of $f(x) = 0$.
   b) Newton's Method will produce a sequence x_1, x_2, ... that does approximate a zero.
   c) $x_2 = x_1$
   d) Newton's Method cannot be used.

Section 3.1

_____ **1.** True, False:
If c is a critical number, then $f'(c) = 0$.

_____ **2.** True, False:
If $f'(c) = 0$, then c is a critical number.

_____ **3.** The critical numbers of $f(x) = 3x^4 + 20x^3 - 36x^2$ are:

      a) 0, 1, 6   b) 0, −1, 6   c) 0, 1, −6   d) 0, −1, −6

_____ **4.** True, False:
The absolute extrema of a continuous function on a closed interval always exist.

_____ **5.** The graph at the right has a
local maximum at $x =$ _____.

    a) 1, 3, and 5
    b) 2
    c) 4
    d) 2 and 4

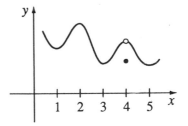

_____ **6.** The absolute maximum value of $f(x) = 6x - x^2$ on $[1, 7]$ is:

    a) 3   b) 9   c) 76   d) −7

_____ **1.** Which of the following is not a hypothesis of Rolle's Theorem?

    a) f is continuous on $[a, b]$.
    b) $f(a) = f(b)$.
    c) $f'(x) \geq 0$ for all $x \in (a, b)$.
    d) f is differentiable on (a, b).

_____ **2.** Yes, No:
Do all the hypotheses for the Mean Value Theorem hold for
$f(x) = 1 - |x|$ on $[-1, 2]$?

_____ **3.** A number c that satisfies the Mean Value Theorem for $f(x) = x^3$ on
$[1, 4]$ is:

    a) $\sqrt{7}$ b) $\sqrt{21}$ c) $\sqrt{63}$ d) 63

_____ **4.** True, False:
If $f'(x) = g'(x)$ for all x then $f(x) = g(x)$.

Section 3.3

_____ **1.** If 4 is a critical number for f and $f'(x) < 0$ for $x < 4$ and $f'(x) > 0$ for $x > 4$, then:

    a) f has a local maximum at 4.
    b) f has a local minimum at 4.
    c) 4 is either a local maximum or minimum, but not enough information is given to decide which.
    d) 4 is neither a local maximum nor local minimum.

_____ **2.** True, False:
$f(x) = 10x - x^2$ is monotonic on $(4, 8)$.

_____ **3.** True, False:
$f(x) = 4x^2 - 8x + 1$ is increasing on $(0, 1)$.

_____ **4.** $f(x) = 12x - x^3$ has a local maximum at:

    a) $x = 0$
    b) $x = 2$
    c) $x = -2$
    d) f does not have a local maximum

_____ **5.** True, False:
If a function is not increasing on an interval, then it is decreasing on the interval.

315

_____ **1.** True, False:
If $f''(c) = 0,$ then c is a point of inflection for f.

_____ **2.** $f(x) = (x^2 - 3)^2$ has points of inflection at:

     a) $0, \sqrt{3}, -\sqrt{3}$ b) $-1, 1$
     c) $0, -1, 1$ d) $\sqrt{3}, -\sqrt{3}$

_____ **3.** $f(x) = x^3 - 2x^2 + x - 1$ has a local maximum at:

     a) 0   b) 1   c) $\frac{1}{3}$ d) f has no local maxima

_____ **4.** What graph has $f'(2) = 3$ and $f''(2) = -4$?

a)

b)

c)

_____ 1. Sometimes, Always, Never:
$$\lim_{x \to \infty} f(x) = \lim_{x \to -\infty} f(x).$$

_____ 2. $\lim\limits_{x \to \infty} \dfrac{\sqrt{x^2+1}}{3x+2} =$

         a) 0   b) 1   c) $\frac{1}{3}$ d) does not exist

_____ 3. True, False:
         If the limits exist, then $\lim\limits_{x \to \infty} (f(x) + g(x)) = \lim\limits_{x \to \infty} f(x) + \lim\limits_{x \to \infty} g(x).$

_____ 4. The horizontal asymptote(s) for $f(x) = \dfrac{|x|}{x+1}$ is(are):

         a) $y = 0$ b) $y = 1$
         c) $y = -1$ d) $y = 1$ and $y = -1$

_____ 5. True, False:
$$\lim_{x \to \infty} \frac{6x^3 + 8x^2 - 9x + 7}{5x^3 - 4x + 30} = \lim_{x \to \infty} \frac{6x^3}{5x^3}.$$

_____ 6. $\lim\limits_{x \to \infty} \dfrac{6 - x^3}{3 + x} =$

         a) ∞ b) $-\infty$ c) 0   d) 2

_____ 1. True, False: The domain is all x for which $f(x)$ is defined.

_____ 2. True, False: The x-intercepts are the values of x for which $f(x) = 0$.

_____ 3. True, False: $f(x)$ is symmetric about the origin if $f(x) = -f(x)$.

_____ 4. True, False: If $f'(x) > 0$ for all x in an interval I then f is increasing on I.

_____ 5. True, False: If $f'(c) = 0$ and $f''(c) < 0$, then c is a local minimum.

_____ 6. True, False: If $f''(x) > 0$ for $x < c$ and $f''(x) < 0$ for $x > c$, then c is a point of inflection.

_____ 7. Which is the graph of $f(x) = x + \frac{1}{x}$?

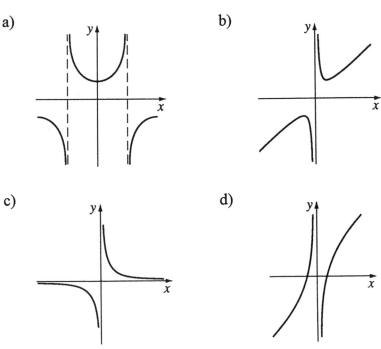

a)

b)

c)

d)

Section 3.7

_____ **1.** The best graph of $f(x) = 3x^5 - 5x^3$ is:

a)

b)

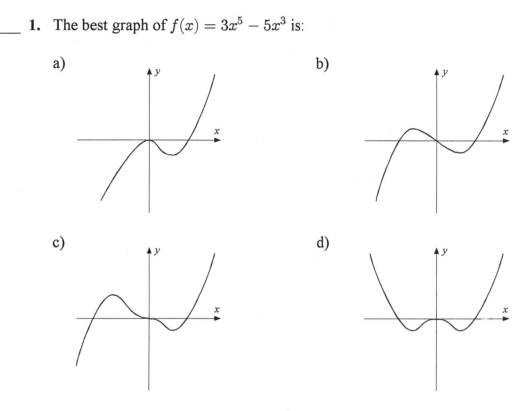

c)

d)

_____ **2.** For the family of functions $f(x) = x^2 - ax$, each function is:

a) a parabola through $(0, 0)$ and $(a, 0)$.
b) is concave up.
c) both a) and b) are correct.

_____ **1.** For two nonnegative numbers, twice the first plus the second is 12. What is the maximum product of two such numbers?

    a) 12

    b) 18

    c) 36

    d) There is no such maximum.

_____ **2.** A carpenter has a 10 foot long board to mark off a triangular area on the floor in the corner of a room. See the figure. What is the function for the triangular area in terms of x used to determine the maximum such area?

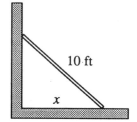

    a) $A = \frac{1}{2}x^2$

    b) $A = \frac{1}{2}(x^2 + 10)$

    c) $A = \frac{1}{2}x\sqrt{100 - x^2}$

    d) $A = \frac{1}{2}x\sqrt{100 - 2x^2}$

Section 3.9

_____ 1. True, False:
Average cost is the derivative of total cost.

_____ 2. True, False:
Marginal revenue is the derivative of total revenue.

_____ 3. Average cost is minimum when:

a) average cost equals total cost
b) average cost equals marginal cost
c) marginal cost equals total cost
d) marginal cost equals zero

_____ 4. Profit is maximum when:

a) total revenue equals total cost
b) marginal revenue equals marginal cost
c) marginal revenue or marginal cost is zero
d) marginal revenue is maximum

_____ 5. Suppose the price (demand function) for an item is $p = 16 - 0.01x$, where x is the number of items. If the cost function is $C(x) = 1000 + 10x + 0.02x^2$, how many items should be made and sold to maximize profits?

a) 50   b) 100   c) 200   d) 250

Section 3.10

_____ 1. True, False:
If $h'(x) = k(x)$, then $k(x)$ is an antiderivative of $h(x)$..

_____ 2. Find $f(x)$ if $f'(x) = 10x^2 + \cos x$.

  a) $20x - \sin x + C$
  b) $20x - \cos x + C$
  c) $\frac{10}{3}x - \cos x + C$
  d) $\frac{10}{3}x^3 - \sin x + C$

_____ 3. Find $f(x)$ if $f'(x) = \frac{1}{x^2}$ and $f(2) = 0$.

  a) $-\frac{1}{x} + \frac{1}{2}$
  b) $-\frac{1}{x} - \frac{1}{2}$
  c) $-\frac{3}{x^3} + \frac{3}{8}$
  d) $-\frac{3}{x^3} - \frac{3}{8}$

_____ 4. True, False:
If $f'(x) = g'(x)$, then $f(x) = g(x)$.

_____ 5. If velocity is $v(t) = 4t + 4$ and $s(1) = 2$, then $s(t) =$

  a) $2t^2 + 4t - 4$
  b) $2t^2 + 4t + 6$
  c) $t^2 + 4$
  d) $t^2 - 4$

Section 4.1

_____ **1.** $\displaystyle\sum_{i=3}^{6} 2i =$

    a) 30   b) 36   c) 40   d) 42

_____ **2.** True, False:

$$\sum_{i=3}^{7} i^2 = \sum_{j=3}^{7} j^2.$$

_____ **3.** $3^3 + 4^3 + 5^3 =$

    a) $\displaystyle\sum_{i=1}^{3} i^3$ b) $\displaystyle\sum_{i=3}^{5} 3i$ c) $\displaystyle\sum_{i=1}^{5} i^3$ d) $\displaystyle\sum_{i=3}^{5} i^3$

_____ **4.** $\displaystyle\sum_{i=1}^{n} (4i^2 + i) =$

    a) $\displaystyle 4\sum_{i=1}^{n} (i^2 + i)$

    b) $\displaystyle 4\sum_{i=1}^{n} i^2 + \sum_{i=1}^{n} i$

    c) $\displaystyle \sum_{i=1}^{n} 5i^2$

    d) $\displaystyle 4\sum_{i=1}^{n} i^2 + 4\sum_{i=1}^{n} i$

Section 4.2

_____ **1.** The norm of the partition $[4, 5]$, $[5, 7]$, $[7, 10]$, $[10, 12]$ of the interval $[4, 12]$ is

a) 1   b) 2   c) 3   d) 8

_____ **2.** Find the sum of the areas of approximating rectangles for the area under $f(x) = 48 - x^2$, between $x = 1$ and $x = 5$. Use the partition $[1, 2]$, $[2, 4]$, $[4, 5]$ and the right endpoints of each subinterval for x_i^*.

a) 131   b) 99   c) 15   d) 192

_____ **3.** True, False:
In general, better approximations to the area under the curve $y = f(x)$ between $x = a$ and $x = b$ are obtained by selection of partitions with smaller norms.

_____ **4.** Selecting midpoints of subintervals for x_i^* to approximate the area under $f(x) = 10x + 5$ between $x = 1$ and $x = 4$ will generate a Riemann sum which is:

a) greater than the actual area
b) less than the actual area
c) equal to the actual area

Section 4.3

_____ 1. Sometimes, Always, or Never:
$\int_a^b f(x)\ dx$ equals the area between $y = f(x)$, the x-axis, $x = a$, and $x = b$.

_____ 2. Sometimes, Always, or Never:
If $f(x)$ is continuous on $[a, b]$, then $\int_a^b f(x)\ dx$ exists.

_____ 3. Using $n = 4$ and midpoints for x_i^*, the Riemann sum for $\int_{-1}^7 x^2\ dx$ is

    a) $\frac{344}{3}$ b) 72   c) 168   d) 112

_____ 4. If A_1 and A_2 are the areas at
the right then $\int_a^b f(x)\ dx =$

    a) $A_1 - A_2$
    b) $-A_1 + A_2$
    c) $A_1 + A_2$
    d) $|A_1 - A_2|$

_____ 5. If $\int_1^3 f(x)\ dx = 10$ and $\int_1^3 g(x)\ dx = 6$, then $\int_1^3 (2f(x) - 3g(x))dx =$

    a) 2   b) 4   c) 18   d) 38

_____ 6. If $f(x) \geq 5$ for all $x \in [2, 6]$ then $\int_2^6 f(x)\ dx \geq$ _____.
(Choose the best answer.)

    a) 4   b) 5   c) 20   d) 30

_____ 7. $\int_3^5 6\ dx =$

    a) 2   b) 8   c) 6   d) 12

_____ 1. If $h(x) - \int_3^x (8t^3 + 2t)dt$, then $h'(x) =$

a) $8t^3 + 2t$ b) $24x^2 + 2$ c) $8x^3 + 2x - 3$ d) $8x^3 + 2x$

_____ 2. $\int f(x)\ dx = F(x)$ means

a) $f'(x) = F(x)$
b) $f(x) = F'(x)$
c) $f(x) = F(b) - F(a)$
d) $f(x) = F(x) + C$

_____ 3. True, or False:
$\int_0^7 2\sqrt[3]{x}\ dx = 2\int_0^7 \sqrt[3]{x}\ dx$

_____ 4. Evaluate $\int_{-1}^1 (x^2 - x^3)dx$

a) $\frac{3}{2}$ b) $\frac{5}{6}$ c) $\frac{1}{2}$ d) $\frac{2}{3}$

_____ 5. $\int (x - \cos x)dx =$

a) $1 - \sin x + C$
b) $1 + \sin x + C$

c) $\frac{x^2}{2} - \sin x + C$

d) $\frac{x^2}{2} + \sin x + C$

_____ **1.** Suppose $\int f(x) = F(x) + C$. Then $\int f(g(x))g'(x)\ dx =$

     a) $F(x)g(x) + C$
     b) $f'(g(x)) + C$
     c) $F(g'(x)) + C$
     d) $F(g(x)) + C$

_____ **2.** What substitution should be made to evaluate $\int \sqrt{x^2 + 2x}\,(x+1)\ dx$?

     a) $u = \sqrt{x^2 + 2x}$
     b) $u = x^2 + 2x$
     c) $u = x + 1$
     d) $u = \sqrt{x^2 + 2x}\,(x+1)$

_____ **3.** What substitution should be made to evaluate $\int \sin^2 x \cos x\ dx$?

     a) $u = \sin x$
     b) $u = \cos x$
     c) $u = \sin x \cos x$
     d) $u = \sin^2 x \cos x$

_____ **4.** Evaluate $\int_0^1 (x^3 + 1)^2 x^2\ dx$.

     a) $\frac{1}{9}$ b) $\frac{1}{3}$ c) $\frac{7}{9}$ d) $\frac{8}{9}$

1. A definite integral for the area shaded at the right is:

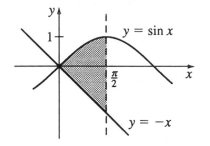

a) $\int_0^{\pi/2} [\sin x - x]dx$

b) $\int_0^{\pi/2} [\sin x + x]dx$

c) $\int_0^1 [\sin x - x]dx$

d) $\int_0^1 [\sin x + x]dx$

2. For the area of the shaded region, which integral form, $\int \dots dx$ or $\int \dots dy$, should be used?

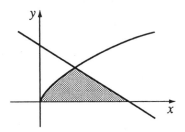

a) $\int \dots dx$

b) $\int \dots dy$

c) Either may be used with equal ease.

3. A definite integral for the area of the region bounded by $y = 2 - x^2$ and $y = x^2$ is:

a) $\int_{-1}^1 [2 - 2x^2]dx$

b) $\int_{-1}^1 [2x^2 - 2]dx$

c) $\int_0^2 [2 - 2x^2]dx$

d) $\int_0^2 [2x^2 - 2]dx$

Section 5.2

_____ 1. Find a definite integral for the volume of a solid with a circular base of radius 4 inches such that the cross section of any slice perpendicular to a certain diameter in the base is a square.

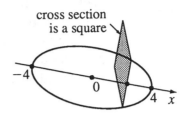

cross section is a square

a) $\int_{-4}^{4}(64 - 4x^2)dx$

b) $\int_{-4}^{4} 4x^2\ dx$

c) $\int_{-4}^{4} 4x^4\ dx$

d) $\int_{-4}^{4}(4 - x^2)^2\ dx$

_____ 2. An integral for the solid obtained by rotating the region at the right about the x-axis is:

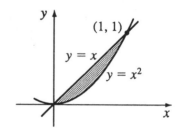

a) $\int_{0}^{1} \pi(x^4 - x^2)dx$

b) $\int_{0}^{1} \pi(x^2 - x^4)dx$

c) $\int_{0}^{1} \pi(x - x^2)dx$

d) $\int_{0}^{1} \pi(x^2 - x)dx$

_____ 3. An integral for the solid obtained by rotating the region at the right about the y-axis is:

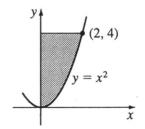

a) $\int_{0}^{4} \pi y\ dy$

b) $\int_{0}^{2} \pi y^2\ dy$

c) $\int_{0}^{4} \pi \sqrt{y}\ dy$

d) $\int_{0}^{2} \pi \sqrt{y}\ dy$

Section 5.3

_____ 1. Which definite integral is the volume of the solid obtained by revolving the region at the right about the y-axis using cylindrical shells?

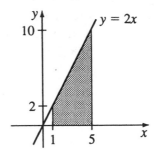

a) $\int_1^5 2\pi x^2 \, dx$

b) $\int_1^5 4\pi x^2 \, dx$

c) $\int_1^5 2\pi x \, dx$

d) $\int_2^{10} 4\pi x \, dx$

_____ 2. Which definite integral is the volume of the solid obtained by revolving the region at the right about the x-axis using cylindrical shells?

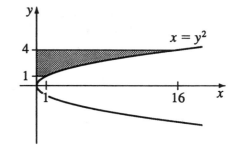

a) $\int_1^{16} 2\pi y^3 \, dy$

b) $\int_1^{16} 2\pi (4 - \sqrt{x}) dy$

c) $\int_1^4 2\pi y^2 \, dy$

d) $\int_1^4 2\pi y^3 \, dy$

Section 5.4

_____ **1.** How much work is done in lifting a 60 pound child 3 feet in the air?

    a) 20 ft-lbs
    b) $20g$ ft-lbs
    c) 180 ft-lbs
    d) $180g$ ft-lbs

_____ **2.** A particle moves along an x-axis from 2 m to 3 m pushed by a force of x^2 newtons. The integral that determines the amount of work done is:

    a) $\int_2^3 gx^3 \, dx$

    b) $\int_2^3 x \, dx$

    c) $\int_2^3 2\pi x^2 \, dx$

    d) $\int_2^3 x^2 \, dx$

1. The average value of $f(x) = 3x^2 + 1$ on the interval $[2, 4]$ is:

   a) 29   b) 66   c) 58   d) 36

2. Which point x_1, x_2, x_3, or x_4 on the graph is the best choice to serve as the point guaranteed by the Mean Value Theorem for Integrals?

   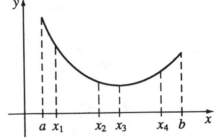

   a) x_1 b) x_2 c) x_3 d) x_4

3. Find a point c for the Mean Value Theorem for Integrals for $f(x) = x^2 - 2x$ on $[2, 5]$.

   a) 18   b) 6   c) $1 + \sqrt{7}$ d) no such c exists

_____ 1. A function f is one-to-one means:

    a) if $x_1 = x_2$, then $f(x_1) = f(x_2)$
    b) if $x_1 \neq x_2$, then $f(x_1) = f(x_2)$
    c) if $x_1 \neq x_2$, then $f(x_1) \neq f(x_2)$
    d) if $f(x_1) \neq f(x_2)$, then $x_1 \neq x_2$

_____ 2. If $f(x) = \sqrt[3]{x+3}$, then $f^{-1}(x) =$

    a) $\frac{1}{\sqrt[3]{x+3}}$ b) $x^3 + 3$ c) $(x+3)^3$ d) $x^3 - 3$

_____ 3. Sometimes, Always, or Never:
    If f is one-to-one and (a, b) is on the graph of $y = f(x)$ then (b, a) is on the graph of $y = f^{-1}(x)$.

_____ 4. If f is one-to-one, differentiable, $f(4) = 7$ and $f'(4) = \frac{1}{2}$ then $(f^{-1})'(7) =$

    a) $\frac{1}{7}$ b) $\frac{1}{4}$ c) $\frac{1}{2}$ d) 2

_____ **1.** Sometimes, Always, or Never: $\lim\limits_{x \to -\infty} a^x = 0$.

_____ **2.** Which is the best graph of $y = 1.5^x$? The graph of $y = 2^x$ is shown as a dashed curve for comparison.

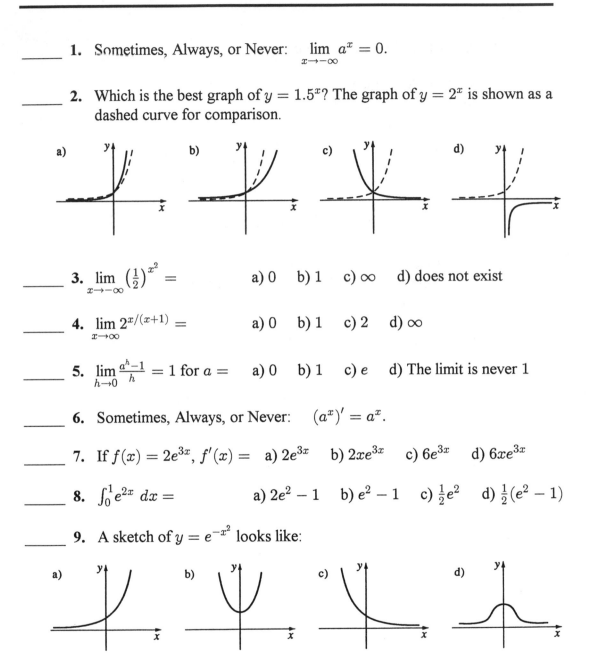

a)   b)   c)   d)

_____ **3.** $\lim\limits_{x \to -\infty} \left(\frac{1}{2}\right)^{x^2} =$ a) 0   b) 1   c) ∞ d) does not exist

_____ **4.** $\lim\limits_{x \to \infty} 2^{x/(x+1)} =$ a) 0   b) 1   c) 2   d) ∞

_____ **5.** $\lim\limits_{h \to 0} \frac{a^h - 1}{h} = 1$ for $a =$ a) 0   b) 1   c) e d) The limit is never 1

_____ **6.** Sometimes, Always, or Never: $(a^x)' = a^x$.

_____ **7.** If $f(x) = 2e^{3x}$, $f'(x) =$ a) $2e^{3x}$ b) $2xe^{3x}$ c) $6e^{3x}$ d) $6xe^{3x}$

_____ **8.** $\int_0^1 e^{2x}\, dx =$ a) $2e^2 - 1$ b) $e^2 - 1$ c) $\frac{1}{2}e^2$ d) $\frac{1}{2}(e^2 - 1)$

_____ **9.** A sketch of $y = e^{-x^2}$ looks like:

a)   b)   c)   d)

Section 6.3

_____ 1. $\log_{27} 9 =$

    a) $\frac{1}{3}$ b) $\frac{2}{3}$ c) $\frac{3}{2}$ d) $\frac{1}{2}$

_____ 2. True, False: $\ln(a + b) = \ln a + \ln b$.

_____ 3. Simplified, $\log_3 9x^3$ is:

    a) $2 + 3\log_3 x$ b) $6\log_3 x$ c) $9\log_3 x$ d) $9 + \log_3 x$

_____ 4. Solve for x: $e^{2x-1} = 10$.

    a) $\frac{1}{2}(1 + 10^e)$ b) $3 + \ln 10$ c) $2 + \ln 10$ d) $\ln\sqrt{10e}$

_____ 5. Solve for x: $\ln(e + x) = 1$.

    a) 0 b) 1 c) $-c$ d) $e^c - e$

_____ **1.** $e^x(e^x)^2 =$

a) e^{x^3} b) e^{3x} c) $3e^x$ d) e^{4x}

_____ **2.** If $f(x) = 2e^{3x}$ then $f'(x) =$

a) $2e^{3x}$ b) $2xe^{3x}$ c) $6e^{3x}$ d) $6xe^{3x}$

_____ **3.** $\int_0^1 e^{2x}\, dx =$

a) $2e^2 - 1$ b) $e^2 - 1$ c) $\frac{1}{2}e^2$ d) $\frac{1}{2}(e^2 - 1)$

_____ **4.** A sketch of $y = e^{-x^2}$ looks like:

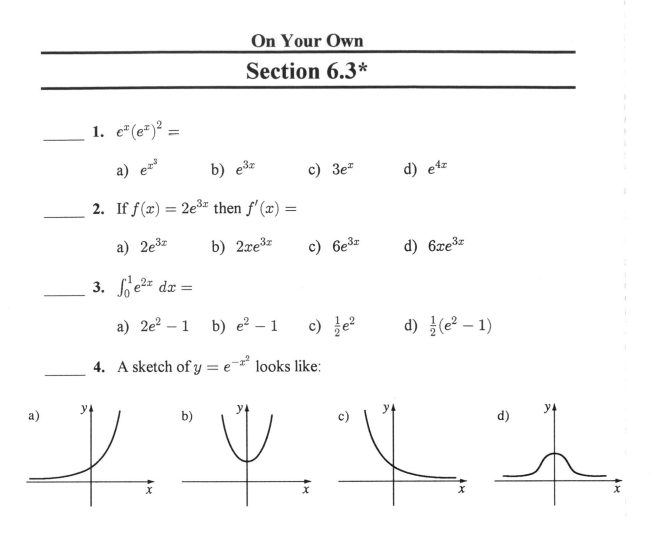

Section 6.4*

_____ **1.** Sometimes, Always, or Never: $\lim\limits_{x \to -\infty} a^x = 0$.

_____ **2.** Which is the best graph of $y = 1.5^x$? The graph of $y = 2^x$ is shown as a dashed curve for comparison.

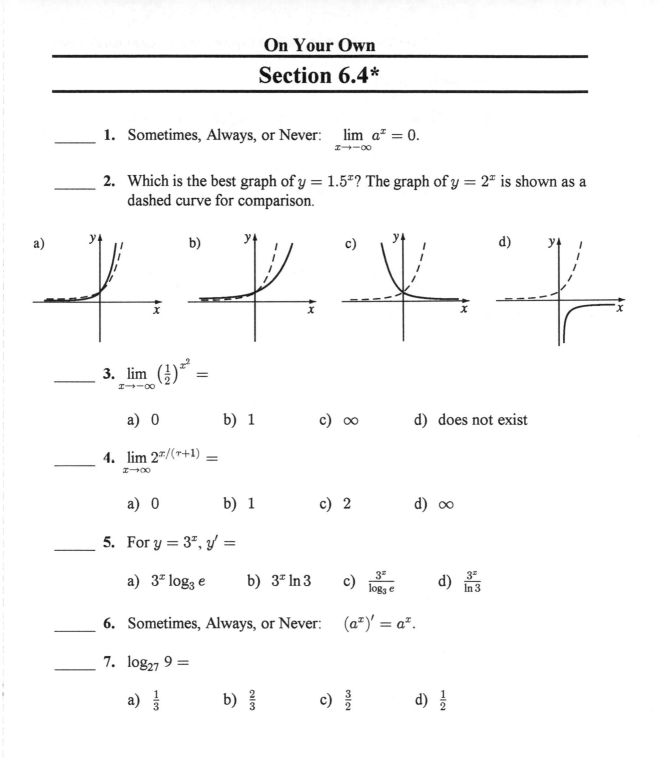

a) b) c) d)

_____ **3.** $\lim\limits_{x \to -\infty} \left(\frac{1}{2}\right)^{x^2} =$

    a) 0   b) 1   c) ∞ d) does not exist

_____ **4.** $\lim\limits_{x \to \infty} 2^{x/(x+1)} =$

    a) 0   b) 1   c) 2   d) ∞

_____ **5.** For $y = 3^x$, $y' =$

    a) $3^x \log_3 e$ b) $3^x \ln 3$ c) $\frac{3^x}{\log_3 e}$ d) $\frac{3^x}{\ln 3}$

_____ **6.** Sometimes, Always, or Never: $(a^x)' = a^x$.

_____ **7.** $\log_{27} 9 =$

    a) $\frac{1}{3}$ b) $\frac{2}{3}$ c) $\frac{3}{2}$ d) $\frac{1}{2}$

_____ **8.** Simplified, $\log_3 9x^3$ is:

     a) $2 + 3\log_3 x$ b) $6\log_3 x$ c) $9\log_3 x$ d) $9 + \log_3 x$

_____ **9.** Solve for x: $e^{2x-1} = 10$.

     a) $\frac{1}{2}(1 + 10^e)$ b) $3 + \ln 10$ c) $2 + \ln 10$ d) $\ln\sqrt{10e}$

_____ **10.** Solve for x: $\ln(e + x) = 1$.

     a) 0 b) 1 c) $-e$ d) $e^e - e$

_____ **11.** $\lim\limits_{x \to 0^+} (1 + x)^{-1/x} =$

     a) e b) $-e$ c) $\frac{1}{e}$ d) $-\frac{1}{e}$

_____ 1. The general solution to $\frac{dy}{dt} = ky$ is:

     a) $y(t) = y(0)e^{kt}$
     b) $y(t) = y(k)e^{t}$
     c) $y(t) = y(t)e^{k}$
     d) $y(t) = e^{y(0)kt}$

_____ 2. A bacteria culture starts with 50 organisms and after 2 hours there are 100. How many will there be after 5 hours?

     a) $50 \ln 5$
     b) $50e^{5}$
     c) $200\sqrt{2}$
     d) 300

_____ **1.** The range of $f(x) = \cos^{-1} x$ is:

a) $\left[-\frac{\pi}{2}, \frac{\pi}{2}\right]$

b) $[0, \pi]$

c) $\left[0, \frac{\pi}{2}\right] \cup \left[\pi, \frac{3\pi}{2}\right]$

d) all reals

_____ **2.** $\sec^{-1}\left(\frac{2}{\sqrt{3}}\right) =$

a) $-\frac{\pi}{3}$ b) $\frac{\pi}{3}$ c) $-\frac{\pi}{6}$ d) $\frac{\pi}{6}$

_____ **3.** For $f(x) = \cos^{-1}(2x)$, $f'(x) =$

a) $\frac{-1}{\sqrt{1-4x^2}}$

b) $\frac{-2}{\sqrt{1-4x^2}}$

c) $\frac{-4}{\sqrt{1-4x^2}}$

d) $\frac{-8}{\sqrt{1-4x^2}}$

_____ **4.** $\int_{-1/2}^{0} \frac{1}{\sqrt{1-4x^2}} \, dx$

a) $\frac{\pi}{4}$ b) $-\frac{\pi}{4}$ c) $\frac{\pi}{3}$ d) $-\frac{\pi}{3}$

Section 6.7

_____ **1.** True, False: $\sinh^2 x + \cosh^2 x = 1$.

_____ **2.** $\cosh 0 =$

    a) 0   b) 1   c) $\frac{e^2}{2}$ d) is not defined

_____ **3.** For $f(x) = \sinh x$, $f''(x) =$

    a) $\cosh^2 x$ b) $\sinh^2 x$ c) $\tanh x$ d) $\sinh x$

_____ **4.** For $f(x) = \tanh^{-1} 2x$, $f'(x) =$

    a) $2(\operatorname{sech}^{-1} 2x)^2$

    b) $2(\operatorname{sech}^2 2x)^{-1}$

    c) $\frac{2}{1-4x^2}$

    d) $\frac{1}{1-4x^2}$

_____ **1.** $\lim\limits_{x \to 5} \frac{x^2 - 25}{x^2 - 9x + 20} =$

      a) 1   b) 10   c) ∞ d) does not exist

_____ **2.** $\lim\limits_{h \to 0^+} \sqrt[3]{x} \ln x =$

      a) 0   b) 1   c) $-\infty$ d) does not exist

_____ **3.** $\lim\limits_{x \to \infty} (\ln x)^{1/x} =$

      a) 0   b) 1   c) e d) ∞

Section 7.1

_____ **1.** Find the correct u and dv for integration by parts of $\int x^2 \cos 2x \; dx$.

a) $u = x^2$, $dv = \cos 2x \; dx$
b) $u = \cos 2x$, $dv = x^2 \; dx$
c) $u = x^2$, $dv = 2x \; dx$
d) $u = x \cos 2x$, $dv = x \; dx$

_____ **2.** Evaluate $\int 16t^3 \ln t \; dt$.

a) $4t^4 + C$
b) $4t^3 \ln t - 16t^3 + C$
c) $4t^4 - t^3 \ln t + C$
d) $4t^4 \ln t - t^4 + C$

_____ **3.** The integration by parts rule comes from which differentiation formula?

a) $(u + v)' = u' + v'$
b) $(uv)' = u'v'$
c) $(uv)' = uv' + u'v$
d) $u \; dv = v \; du$

_____ 1. By the methods of trigonometric integrals, $\int \sin^3 x \cos^2 x \, dx$ may be rewritten as:

a) $\int \sin^3 x (1 - \sin^2 x) dx$

b) $\int (1 - \cos^2 x)^{3/2} \cos^2 x \, dx$

c) $\int (\sin^3 x)(\cos x) \cos x \, dx$

d) $\int (1 - \cos^2 x) \cos^2 x \sin x \, dx$

_____ 2. Using $\tan^2 x = \sec^2 x - 1$, $\int \tan^3 x \, dx$

a) $\frac{3}{2} \ln |\sec^2 x - 1| + C$

b) $\frac{1}{2} \tan^2 x - \ln |\sec x| + C$

c) $\frac{1}{2} \tan^2 x - \frac{1}{4} \sec^2 x + C$

d) $\frac{1}{4} \tan^2 x + \ln |\sec x| + C$

_____ **1.** What trigonometric substitution should be made for $\int \frac{x^2}{\sqrt{x^2-25}} \, dx$?

    a) $x = 5\sin\theta$
    b) $x = 5\tan\theta$
    c) $x = 5\sec\theta$
    d) No such substitution will evaluate the integral

_____ **2.** Rewrite $\int \frac{\sqrt{10-x^2}}{x} \, dx$ using a trigonometric substitution.

    a) $\int \frac{\cos\theta}{\sqrt{10}\sin\theta} \, d\theta$

    b) $\int \frac{1}{\sqrt{10}} \sin\theta \cos^2\theta \, d\theta$

    c) $\int \frac{\sqrt{10}\cos^2\theta}{\sin\theta} \, d\theta$

    d) $\int \sqrt{10}\sin\theta \cos\theta \, d\theta$

_____ **3.** Is a trigonometric substitution necessary for $\int \frac{x}{\sqrt{x^2+10}} \, dx$?

    a) Yes   b) No

Section 7.4

_____ **1.** What is the partial fraction form of $\frac{2x+3}{(x-2)(x+2)}$?

     a) $\frac{Ax}{x-2} + \frac{B}{x+2}$

     b) $\frac{A}{x-2} + \frac{B}{x+2}$

     c) $\frac{Ax}{x-2} + \frac{Bx}{x+2}$

     d) $\frac{Ax+B}{x-2} + \frac{Cx+D}{x+2}$

_____ **2.** What is the partial fraction form of $\frac{2x+3}{(x^2+1)^2}$?

     a) $\frac{Ax+B}{(x^2+1)^2}$

     b) $\frac{A}{x^2+1} + \frac{B}{(x^2+1)^2}$

     c) $\frac{A}{x^2+1} + \frac{Bx+C}{(x^2+1)^2}$

     d) $\frac{Ax+B}{x^2+1} + \frac{Cx+D}{(x^2+1)^2}$

_____ **3.** Evaluate $\int \frac{2}{(x+1)(x+2)}\ dx$, using partial fractions.

     a) $2\ln|x+1| - 2\ln|x+2| + C$
     b) $\ln|x+1| + 3\ln|x+2| + C$
     c) $3\ln|x+1| - \ln|x+2| + C$
     d) $\ln|x+1| + \ln|x+2| + C$

Section 7.5

_____ 1. What rationalizing substitution should be made for $\int \frac{\sqrt[3]{x}+2}{\sqrt[3]{x}+1}\, dx$?

a) $u = x$
b) $u = \sqrt[3]{x}$
c) $u = \sqrt[3]{x} + 2$
d) $u = \sqrt[3]{x} + 1$

_____ 2. What rationalizing substitution should be made for $\int \frac{1}{\sqrt{x}+\sqrt[5]{x}}\, dx$?

a) $u = x$
b) $u = \sqrt{x}$
c) $u = \sqrt[5]{x}$
d) $u = \sqrt[10]{x}$

_____ 3. What integral results from the substitution $t = \tan \frac{x}{2}$ in $\int \frac{\sin x}{1-\cos x}\, dx$?

a) $\int \frac{2}{t(1+t^2)}\, dt$

b) $\int \frac{2t^2}{1-t^2}\, dt$

c) $\int \frac{4t^4}{1+t^2}\, dt$

d) $\int \frac{2t^2}{(1+t^2)^2}\, dt$

_____ **1.** True, False:

$\int x^2 \sqrt{x^2 - 4} \; dx$ should be evaluated using integration by parts.

_____ **2.** True, False:

A straight (simple) substitution may be used to evaluate $\int \frac{2 + \ln x}{x} \; dx$.

_____ **3.** True, False:

Partial fractions may be used for $\int \frac{1}{\sqrt{x+2}\sqrt{x+3}} \; dx$.

_____ **4.** True, False:

$\int \frac{\sqrt{x^2+1}}{x^3} \; dx$ may be solved using a trigonometric substitution.

_____ **5.** Sometimes, Always, or Never:

The antiderivative of an elementary function is elementary.

Section 7.7

_____ **1.** What is the number of the integral formula in your text's Table of Integrals that may be used to evaluate $\int \frac{dx}{x\sqrt{2-x^2}}$?

a) #35   b) #36   c) #18   d) #24

_____ **2.** What is the number of the integral formula in your text's Table of Integrals that may be used to evaluate $\int \frac{x^2 dx}{(x^2+2x+5)^{3/2}}$?

a) #37   b) #38   c) #26   d) #46

_____ 1. If $f(x)$ is integrable and concave upward on $[a, b]$, then an estimate of $\int_a^b f(x)\, dx$ using the Trapezoidal Rule will always be:

   a) too large
   b) too small
   c) exact
   d) within $\frac{b-a}{12n}$ of the exact value

_____ 2. An estimate of $\int_{-1}^3 x^4\, dx$ using Simpson's Rule with $n = 4$ gives:

   a) $\frac{242}{5}$ b) $\frac{148}{3}$ c) $\frac{152}{3}$ d) $\frac{244}{5}$

_____ 3. Using $\frac{M(b-a)^3}{24n^2}$ find the maximum error in estimating $\int_{-1}^3 x^4\, dx$ with the Midpoint Rule and 8 subintervals.

   a) $\frac{1}{2}$ b) 9 c) $\frac{9}{2}$ d) $\frac{9}{8}$

_____ **1.** True, False:

$\int_{-2}^{3} |x| \, dx$ is an improper integral.

_____ **2.** By definition the improper integral $\int_{1}^{e} \frac{1}{x \ln x} \, dx =$

a) $\lim\limits_{t \to 1^+} \int_{t}^{e} \frac{1}{x \ln x} \, dx$

b) $\lim\limits_{t \to e^-} \int_{1}^{t} \frac{1}{x \ln x} \, dx$

c) $\int_{1}^{2} \frac{1}{x \ln x} \, dx + \int_{2}^{e} \frac{1}{x \ln x} \, dx$

d) It is not an improper integral

_____ **3.** $\int_{0}^{2} \frac{x-2}{\sqrt{4x - x^2}} \, dx =$

a) 0   b) 2   c) -2 d) does not exist

_____ **4.** Suppose we know $\int_{1}^{\infty} f(x) \, dx$ diverges and that $f(x) \geq g(x) \geq 0$ for all $x \geq 1$. What conclusion can be made about $\int_{1}^{\infty} g(x) \, dx$?

a) it converges to the value of $\int_{1}^{\infty} f(x) \, dx$
b) it converges, but we don't know to what value
c) it diverges
d) no conclusion (it could diverge or converge)

_____ 1. A differential equation is separable if it can be written in the form:

a) $\frac{dy}{dx} = f(x) + g(y)$ b) $\frac{dy}{dx} = f(g(x, y))$

c) $\frac{dy}{dx} = \frac{f(x)}{g(y)}$ d) $\frac{dy}{dx} = f(x)^{g(y)}$

_____ 2. $(y''')^2 + 6xy - 2y' + 7x = 0$ has order:

a) 1   b) 2   c) 3   d) 4

_____ 3. The general solution to $\frac{dy}{dx} = \frac{1}{xy}$ (for x, $y > 0$) is:

a) $y = \ln x + C$ b) $y = \sqrt{\ln x + C}$

c) $y = \ln(\ln x + C)$ d) $y = \sqrt{\ln x^2 + C}$

_____ 4. The solution to $y' = -y^2$ with $y(1) = \frac{1}{3}$ is:

a) $y = \frac{1}{x+2}$ b) $y = \ln x + 3$

c) $y = \frac{1}{2x^2} - \frac{1}{6}$ d) $y = \frac{1}{2x} - \frac{1}{6}$

_____ 5. The directional field at the right is for:

a) $y' = x - y$
b) $y' = xy$
c) $y' = x + y$
d) $y' = \frac{x}{y}$

Section 8.2

_____ **1.** A definite integral for the length of $y = x^3$, $1 \le x \le 2$ is:

a) $\int_1^2 \sqrt{1 + x^3}\ dx$

b) $\int_1^2 \sqrt{1 + x^6}\ dx$

c) $\int_1^2 \sqrt{1 + 9x^4}\ dx$

d) $\int_1^2 \sqrt{1 + 3x^2}\ dx$

_____ **2.** The arc length function for $y = x^2$, $1 \le x \le 3$ is $s(x) =$

a) $\int_1^x \sqrt{1 + 4x^2}\ dx$

b) $\int_1^x \sqrt{1 + 4t^2}\ dt$

c) $\int_1^x \sqrt{1 + t^4}\ dt$

d) $\int_1^x \sqrt{1 + 4t^4}\ dt$

Section 8.3

_____ 1. An integral for the surface area obtained by rotating $y = \sin x + \cos x$, $0 \le x \le \frac{\pi}{2}$, about the x-axis is:

a) $\int_0^{\pi/2} 2\pi(\sin x + \cos x)\sqrt{2 - 2\sin x \cos x}\ dx$

b) $\int_0^{\pi/2} 2\pi(\sin x + \cos x)\sqrt{1 + \sin^2 x - \cos^2 x}\ dx$

c) $\int_0^{\pi/2} 2\pi(\cos x - \sin x)\sqrt{1 + \sin^2 x - \cos^2 x}\ dx$

d) $\int_0^{\pi/2} 2\pi(\cos x - \sin x)\sqrt{2 - 2\sin x \cos x}\ dx$

_____ 2. A hollow cylinder of radius 3 in. and height 4 in. is a surface of revolution. Its round surface area may be expressed as:

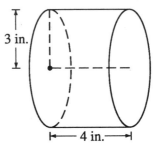

a) $\int_0^3 2\pi(4)\sqrt{1 + 0}\ dx$

b) $\int_0^3 2\pi(3)\sqrt{1 + 0}\ dx$

c) $\int_0^4 2\pi(3)\sqrt{1 + 0}\ dx$

d) $\int_0^4 2\pi(4)\sqrt{1 + 0}\ dx$

Section 8.4

_____ 1. The x-coordinate of the center of mass of the region bounded by $y = \frac{1}{x}$, $x = 1$, $x = 2$, $y = 0$ is:

    a) $\ln 2$ b) $\frac{1}{\ln 2}$ c) 1 d) $2\ln 2$

_____ 2. The y-coordinate of the center of mass of the region bounded by $y = \sqrt{x - 1}$, $x = 1$, $x = 5$, $y = 0$ is:

    a) $\frac{3}{14}$ b) $\frac{3}{4}$ c) $\frac{14}{3}$ d) 4

_____ 3. True, False:

$$\overline{x} = \frac{M_y}{\rho A} \text{ and } \overline{y} = \frac{M_x}{\rho A}.$$

Section 8.5

_____ **1.** Find a definite integral for the hydrostatic force exerted by a liquid with density 800 kg/m³ on the semicircular plate in the center of the figure.

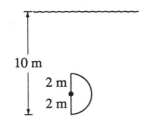

a) $\int_{-2}^{2} 800(9.8)[8 - y]\left[2\sqrt{4 - y^2}\right] dy$

b) $\int_{-2}^{2} 800(9.8)[8 - y]\sqrt{4 - y^2}\ dy$

c) $\int_{-2}^{2} 800(9.8)[10 - y]\left[2\sqrt{4 - y^2}\right] dy$

d) $\int_{-2}^{2} 800(9.8)[10 - y]\sqrt{4 - y^2}\ dy$

Section 8.6

_____ 1. True, False:

In many applications of definite integrals, the integral is used to compute the total amount of a varying quantity.

_____ 2. Find a definite integral for the consumer surplus if 180 units are available and the demand function is $p(x) = 3000 - 10x$.

a) $\int_0^{180}(3000 - 10x)dx$

b) $\int_0^{180}(1800 - 10x)dx$

c) $\int_0^{180}(1200 - 10x)dx$

d) $\int_0^{180}(180 - 10x)dx$

_____ 3. Find a definite integral for the present value of an income stream of $5000 per year starting next year for 4 years if interest is compounded continuously at 8%.

a) $\int_1^5 5000e^{-0.08t}\ dt$

b) $\int_1^5 5000te^{-0.08t}\ dt$

c) $\int_0^4 5000e^{-0.08t}\ dt$

d) $\int_0^4 5000te^{-0.08t}\ dt$

_____ **1.** The graph of $x - 2 + 3t$, $y = 4 - t$ is a:

    a) circle   b) ellipse

    c) line   d) parabola

_____ **2.** The best graph of $x = \cos t$, $y = \sin^2 t$ is:

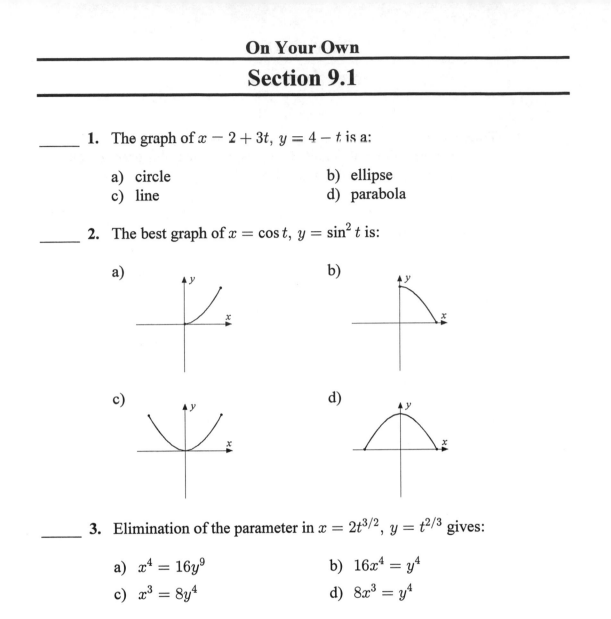

    a)   b)

    c)   d)

_____ **3.** Elimination of the parameter in $x = 2t^{3/2}$, $y = t^{2/3}$ gives:

    a) $x^4 = 16y^9$ b) $16x^4 = y^4$

    c) $x^3 = 8y^4$ d) $8x^3 = y^4$

_____ **1.** Find $\frac{dy}{dx}$ if $x = \sqrt{t}$, $y = \sin 2t$.

     a) $\frac{4\cos t}{\sqrt{t}}$ b) $\frac{\cos 2t}{\sqrt{t}}$ c) $\frac{\cos 2t}{2\sqrt{t}}$ d) $4\sqrt{t}\cos 2t$

_____ **2.** Find $\frac{d^2 y}{dx^2}$ for $x = 3t^2 + 1$, $y = t^6 + 6t^5$.

     a) $t^4 + 5t^3$ b) $4t^3 + 15t^2$

     c) $\frac{2}{3}t^3 + \frac{5}{2}t$ d) $t^3 + \frac{1}{2}t^2$

_____ **3.** The slope of the tangent line at the point where $t = \frac{\pi}{6}$ to the curve $y = \sin 2x$, $x = \cos 3x$ is:

     a) $\frac{1}{3}$ b) $-\frac{1}{3}$ c)   3   d) -3

_____ **4.** A definite integral for the area under the curve described by $x = t^2 + 1$, $y = 2t$, $0 \le t \le 1$ is:

     a) $\int_0^1 (2t^3 + 2t)dt$ b) $\int_0^1 4t^2\ dt$

     c) $\int_0^1 (2t^2 + 2)dt$ d) $\int_0^1 4t\ dt$

Section 9.3

_____ 1. The length of the curve given by $x = 3t^2 + 2$, $y = 2t^3$, $t \in [0, 1]$ is:

a) $4\sqrt{2} - 2$

b) $8\sqrt{2} - 1$

c) $\frac{2}{3}\left(2\sqrt{2} - 1\right)$

d) $\sqrt{2} - 1$

_____ 2. Find a definite integral for the area of the surface of revolution about the x-axis obtained by rotating the curve $y = t^2$, $x = 1 + 3t$, $0 \le t \le 2$.

a) $\int_0^2 2\pi t^2 \sqrt{t^4 + 9t^2 + 6t + 1}\ dt$

b) $\int_0^2 2\pi t^2 \sqrt{4t^2 + 9}\ dt$

c) $\int_0^2 2\pi(2t) \sqrt{t^4 + 9t^2 + 6t + 1}\ dt$

d) $\int_0^2 2\pi(2t) \sqrt{4t^4 + 9}\ dt$

_____ **1.** The polar coordinates of the
point plotted at the right are:

a) $\left(-2, \frac{\pi}{4}\right)$

b) $\left(-2, \frac{3\pi}{4}\right)$

c) $\left(2, -\frac{\pi}{4}\right)$

d) $\left(2, \frac{3\pi}{4}\right)$

_____ **2.** Polar coordinates of the point with rectangular coordinates $(5, 5)$ are:

a) $(25, 0)$ 　　　　　　b) $\left(5, \frac{\pi}{4}\right)$

c) $\left(5\sqrt{2}, \frac{\pi}{4}\right)$ 　　　d) $\left(50, -\frac{\pi}{4}\right)$

_____ **3.** Rectangular coordinates of the point with polar coordinates $\left(-1, \frac{3\pi}{2}\right)$
are:

a) $(-1, 0)$ 　　　　　b) $(0, 1)$

c) $(0, -1)$ 　　　　　d) $(1, 0)$

_____ **4.** The graph of $\theta = 2$ in polar coordinates is a:

a) circle 　　b) line 　　c) spiral 　　d) 3-leaved rose

_____ **5.** The slope of the tangent line to $r = \cos\theta$ at $\theta = \frac{\pi}{3}$ is:

a) $\sqrt{3}$ 　　b) $\frac{1}{\sqrt{3}}$ 　　c) $-\sqrt{3}$ 　　d) $-\frac{1}{\sqrt{3}}$

_____ **1.** The area of the region bounded by $\theta = \frac{\pi}{3}$, $\theta = \frac{\pi}{4}$, and $r = \sec\theta$ is:

a) $\frac{1}{2}(\sqrt{3}-1)$ b) $\sqrt{3}$

c) $2(\sqrt{3}-1)$ d) $2\sqrt{3}$

_____ **2.** The area of the shaded region is given by:

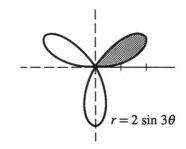

$r = 2\sin 3\theta$

a) $\int_0^{\pi/2} \sin 3\theta \, d\theta$

b) $\int_0^{\pi/2} 2\sin^2 3\theta \, d\theta$

c) $\int_0^{\pi/3} \sin 3\theta \, d\theta$

d) $\int_0^{\pi/3} 2\sin^2 3\theta \, d\theta$

_____ **3.** The length of the arc $r = e^\theta$ for $0 \le \theta \le \pi$ is:

a) $e^\pi - 1$ b) $2(e^\pi - 1)$

c) $\sqrt{2}(e^\pi - 1)$ d) $2\sqrt{2}(e^\pi - 1)$

_____ 1. The figure at the right shows one point P on a conic and the distances of P to the focus and the directrix of a conic. What type of conic is it?

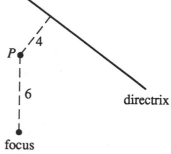

a) parabola
b) ellipse
c) hyperbola
d) not enough information is provided to answer the question

_____ 2. The polar equation of the conic with eccentricity 3 and directrix $x = -7$ is:

a) $r = \frac{21}{1+3\cos\theta}$

b) $r = \frac{21}{1-3\cos\theta}$

c) $r = \frac{21}{1+3\sin\theta}$

d) $r = \frac{21}{1-3\sin\theta}$

_____ 3. The directrix of the conic given by $r = \frac{6}{2+10\sin\theta}$ is:

a) $x = \frac{5}{3}$ b) $x = \frac{3}{5}$ c) $y = \frac{5}{3}$ d) $y = \frac{3}{5}$

_____ 4. The polar form for the graph at the right is:

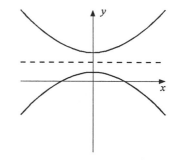

a) $r = \frac{ed}{1+e\cos\theta}$

b) $r = \frac{ed}{1-e\cos\theta}$

c) $r = \frac{ed}{1+e\sin\theta}$

d) $r = \frac{ed}{1-e\sin\theta}$

365

_____ **1.** $\lim\limits_{n\to\infty} \frac{n^2+3n}{2n^2+n+1} =$

a) 0   b) $\frac{1}{2}$ c) 1   d) ∞

_____ **2.** Sometimes, Always, or Never:
If $a_n \geq b_n \geq 0$ and $\{b_n\}$ diverges, then $\{a_n\}$ diverges.

_____ **3.** Sometimes, Always, or Never:
If $a_n \geq b_n \geq c_n$ and both $\{a_n\}$ and $\{c_n\}$ converge, then $\{b_n\}$ converges.

_____ **4.** Sometimes, Always, or Never:
If $\{a_n\}$ is increasing and bounded above, then $\{a_n\}$ converges.

_____ **5.** True, False:
$a_n = \frac{(-1)^n}{n^2}$ is monotonic.

_____ **6.** $\lim\limits_{n\to\infty} \frac{\arctan n}{2} =$

a) $\frac{\pi}{4}$ b) $\frac{\pi}{2}$ c) π d) does not exist

_____ **7.** The fourth term of $\{a_n\}$ defined by $a_1 = 3$, and $a_{n+1} = \frac{2}{3}a_n$, $n = 1, 2,$ $3, \ldots,$ is:

a) $\frac{16}{81}$ b) $\frac{16}{27}$ c) $\frac{8}{27}$ d) $\frac{8}{9}$

_____ **1.** True, False:

$$\sum_{n=1}^{\infty} a_n \text{ converges if } \lim_{n \to \infty} a_n = 0.$$

_____ **2.** True, False:

If $\sum_{n=1}^{\infty} a_n$ converges and $\sum_{n=1}^{\infty} b_n$ converges, then $\sum_{n=1}^{\infty} (a_n - b_n)$ converges.

_____ **3.** The harmonic series is:

a) $1 + 2 + 3 + 4 + \ldots$

b) $1 + \frac{1}{2} + \frac{1}{3} + \frac{1}{4} + \ldots$

c) $1 + \frac{1}{2} + \frac{1}{4} + \frac{1}{8} + \ldots$

d) $1 - \frac{1}{2} + \frac{1}{4} - \frac{1}{8} + \ldots$

_____ **4.** True, False:

If $\sum_{n=1}^{\infty} \frac{n}{n+1}$ converges.

_____ **5.** $\sum_{n=1}^{\infty} 2\left(\frac{1}{4}\right)^n$ converges to:

a) $\frac{9}{4}$ b) 2 c) $\frac{2}{3}$ d) the series diverges

_____ **6.** True, False:

$-3 + 1 - \frac{1}{3} + \frac{1}{9} - \frac{1}{27} + \ldots$ is a geometric series.

_____ 1. For what values of p does the series $\sum_{n=1}^{\infty} \frac{1}{(n^2)^p}$ converge?

  a) $p > -\frac{1}{2}$ b) $p < -\frac{1}{2}$ c) $p > \frac{1}{2}$ d) $p < \frac{1}{2}$

_____ 2. True, False:
If $f(x)$ is continuous and decreasing, $f(n) = a_n$ for all $n = 1, 2, 3 \ldots,$
and $\int_1^{\infty} f(x)dx = M$, then $\sum_{n=1}^{\infty} a_n = M$.

_____ 3. Does $\sum_{n=1}^{\infty} \frac{1}{n^{2/3}}$ converge?

_____ 4. Does $\sum_{n=1}^{\infty} \frac{n+2}{(n^2+4n+1)^2}$ converge?

_____ 5. For $s = \sum_{n=1}^{\infty} \frac{1}{n^3}$, an upper bound estimate for $s - s_6$ (where s_6 is the sixth partial sum) is:

  a) $\int_1^{\infty} x^{-3}\,dx$ b) $\int_6^{\infty} x^{-3}\,dx$

  c) $\int_7^{\infty} x^{-3}\,dx$ d) the series does not converge

Section 10.4

_____ **1.** Sometimes, Always, or Never:

If $0 \le a_n \le b_n$ for all n and $\sum_{n=1}^{\infty} a_n$ diverges, then $\sum_{n=1}^{\infty} b_n$ converges.

_____ **2.** Does $\sum_{n=1}^{\infty} \frac{n+1}{n^3}$ converge?

_____ **3.** Does $\sum_{n=1}^{\infty} \frac{\sqrt{n}+\sqrt[3]{n}}{n^{2/3}+n^{3/2}+1}$ converge?

_____ **4.** Does $\sum_{n=1}^{\infty} \frac{\cos^2(2^n)}{2^n}$ converge?

_____ **5.** $s = \sum_{n=1}^{\infty} \frac{1}{n \cdot 2^n}$ converges by the Comparison Test, comparing it to $\sum_{n=1}^{\infty} \frac{1}{2^n}$.
Using this information, make an estimate of the difference between s and its third partial sum.

a) $\frac{15}{16}$ b) $\frac{7}{8}$ c) $\frac{31}{32}$ d) 1

_____ **1.** Does $\sum_{n=1}^{\infty} \frac{(-1)^{n+1} \ln n}{n^2}$ converge?

_____ **2.** Does $\sum_{n=1}^{\infty} \frac{(-1)^n}{\sqrt[4]{n+1}}$ converge?

_____ **3.** For what value of n is the nth partial sum within 0.001 of the value of $\sum_{n=1}^{\infty} \frac{(-1)^n}{2^n}$?

    a) $n = 4$ b) $n = 6$ c) $n = 8$ d) $n = 10$

_____ **4.** True, False:

The Alternating Series Test may be applied to determine the convergence of $\sum_{n=1}^{\infty} \frac{1+(-1)^n}{2n^2}$.

Section 10.6

_____ **1.** True, False:

If $\sum_{n=1}^{\infty} a_n$ converges absolutely, then it converges conditionally.

_____ **2.** True, False:

If $\lim_{n \to \infty} \left| \frac{a_n}{a_{n+1}} \right| = 3,$ then $\sum_{n=1}^{\infty} a_n$ converge absolutely.

_____ **3.** True, False:

Every series must do one of these: converge absolutely, converge conditionally, or diverge.

_____ **4.** The series $\sum_{n=1}^{\infty} 2^{-n} n!$ _____.

  a) diverges
  b) converges absolutely
  c) converges conditionally
  d) converges, but not absolutely or conditionally

_____ **5.** The series $\sum_{n=1}^{\infty} \frac{(-5)^{n+1}}{n^n}$ _____.

  a) diverges
  b) converges absolutely
  c) converges conditionally
  d) converges, but not absolutely or conditionally

_____ **1.** True, False:

$$\sum_{n=2}^{\infty} \frac{1}{(\ln n)^n} \text{ converges.}$$

_____ **2.** True, False:

$$\sum_{n=1}^{\infty} \frac{6}{7n+8} \text{ converges.}$$

_____ **3.** True, False:

$$\sum_{n=1}^{\infty} \frac{(-1)^n \sqrt{n}}{n+3} \text{ converges.}$$

_____ **4.** True, False:

$$\sum_{n=1}^{\infty} \frac{e^n}{n!} \text{ converges.}$$

_____ 1. Sometimes, Always, Never:

The interval of convergence of a power series $\sum\limits_{n=0}^{\infty} a_n (x - c)^n$ is an open interval $(c - R, c + R)$. (When $R = 0$ we mean $\{0\}$ and when $R = \infty$ we mean $(-\infty, \infty)$.)

_____ 2. True, False:

If a number p is in the interval of convergence of $\sum\limits_{n=0}^{\infty} a_n x^n$ then so is the number $\frac{p}{2}$.

_____ 3. For $f(x) = \sum\limits_{n=0}^{\infty} \frac{(x-1)^n}{3^n}$, $f(3) =$

a) 0   b) 2   c) 3   d) $f(3)$ does not exist

_____ 4. The interval of convergence of $\sum\limits_{n=1}^{\infty} \frac{x^n}{\sqrt{n}}$ is:

a) $[-1, 1]$ b) $[-1, 1)$ c) $(-1, 1]$ d) $(-1, 1)$

_____ 5. The radius of convergence of $\sum\limits_{n=0}^{\infty} \frac{n(x-5)^n}{3^n}$ is:

a) $\frac{1}{3}$ b) 1   c) 3   d) ∞

_____ **1.** Given that $e^x = \sum_{n=0}^{\infty} \frac{x^n}{n!}$, a power series for xe^{x^2} is:

a) $\sum_{n=0}^{\infty} \frac{x^2}{n!}$

b) $\sum_{n=0}^{\infty} \frac{x^{2n}}{n!}$

c) $\sum_{n=0}^{\infty} \frac{x^{2n+1}}{n!}$

d) $\sum_{n=0}^{\infty} \frac{x^{2n}}{(n+1)!}$

_____ **2.** For $f(x) = \sum_{n=0}^{\infty} \frac{x^{2n}}{n!}$, $f'(x) =$

a) $\sum_{n=1}^{\infty} x^{2n-1}$

b) $\sum_{n=1}^{\infty} 2x^{2n-1}$

c) $\sum_{n=1}^{\infty} 2^n x^{2n-1}$

d) $\sum_{n=1}^{\infty} (2n-1)x^{2n-1}$

_____ **3.** Using $\frac{1}{1-x} = \sum_{n=0}^{\infty} x^n$, $\int \frac{x}{1-x^2} \, dx =$

a) $\sum_{n=0}^{\infty} \frac{x^{2n}}{2n}$

b) $\sum_{n=0}^{\infty} \frac{x^{2n+1}}{2n+1}$

c) $\sum_{n=0}^{\infty} \frac{x^{2n+2}}{2n+2}$

d) $\sum_{n=0}^{\infty} x^{2n+1}$

_____ **1.** Given the Taylor series $e^x = \sum_{n=0}^{\infty} \frac{x^n}{n!}$, a Taylor series for $e^{x/2}$ is:

a) $\sum_{n=0}^{\infty} \frac{2^n x^n}{n!}$
   b) $\sum_{n=0}^{\infty} \frac{2x^n}{n!}$
   c) $\sum_{n=0}^{\infty} \frac{x^n}{2^n n!}$
   d) $\sum_{n=0}^{\infty} \frac{x^n}{2\,n!}$

_____ **2.** The nth term in the Taylor series about $x = 1$ for $f(x) = x^{-2}$ is:

a) $(-1)^{n+1}(n+1)!(x-1)^n$ b) $(-1)^n n!(x-1)^n$

c) $(-1)^{n+1}(n+1)(x-1)^n$ d) $(-1)^n n(x-1)^n$

_____ **3.** The Taylor polynomial of degree 3 for $f(x) = x(\ln x - 1)$ about $x = 1$ is $T_3(x) =$

a) $-1 + \frac{x^2}{2} - \frac{x^3}{6}$ b) $-1 + x - \frac{x^2}{2} + \frac{x^3}{6}$

c) $-1 + x - x^2 + x^3$ d) $-1 - x + x^2 - x^3$

_____ **4.** True, False:
If $T_n(x)$ is the nth Taylor polynomial for $f(x)$ at $x = c$, then $T_n^{(k)}(c) = f^{(k)}(c)$ for $k = 0, 1, \ldots, n$.

_____ **5.** Taylor's formula with $n = 3$ for $f(x) = \sqrt{x}$ about $c = 9$ is:

a) $-\frac{5(x-9)^3}{128 z^{5/2}}$
  b) $-\frac{5(x-9)^4}{128 z^{7/2}}$
  c) $-\frac{15(x-9)^3}{16 z^{5/2}}$
  d) $-\frac{15(x-9)^4}{16 z^{7/2}}$

_____ 1. $\left(\begin{smallmatrix} \frac{1}{4} \\ 3 \end{smallmatrix}\right) -$

      a) $\frac{5}{16}$ b) $-\frac{1}{16}$ c) $\frac{5}{8}$ d) $-\frac{5}{8}$

_____ 2. $\sum_{n=0}^{\infty} \left(\begin{smallmatrix} \frac{2}{3} \\ n \end{smallmatrix}\right) x^n$ is the binomial series for:

      a) $(1+x)^{2/3}$ b) $(1+x)^{-2/3}$

      c) $(1+x)^{3/2}$ d) $(1+x)^{-3/2}$

_____ 3. Using a binomial series, the Maclaurin series for $\frac{1}{1+x^2}$ is:

      a) $\sum_{n=0}^{\infty} \left(\begin{smallmatrix} 1 \\ n \end{smallmatrix}\right) x^n$ b) $\sum_{n=0}^{\infty} \left(\begin{smallmatrix} -1 \\ n \end{smallmatrix}\right) x^n$

      c) $\sum_{n=0}^{\infty} \left(\begin{smallmatrix} 1 \\ n \end{smallmatrix}\right) x^{2n}$ d) $\sum_{n=0}^{\infty} \left(\begin{smallmatrix} -1 \\ n \end{smallmatrix}\right) x^{2n}$

_____ 1. If the Maclaurin polynomial of degree 2 for $f(x) = e^x$ is used to approximate $e^{0.2}$ then an estimate for the error with $0 < z < 0.2$ is:

a) $\frac{1}{24}e^z$ b) $\frac{1}{6}e^z$ c) $\frac{1}{2}e^z$ d) e^z

_____ 2. Let $f(x) = e^{2x}$, $c = 0$. Use Taylor's formula to estimate the accuracy of $f(x) \approx T_4(x)$ when $-1 \le x \le 1$.

a) $\frac{4e^2}{15}$ b) $\frac{4}{15}$ c) $\frac{e^2}{3}$ d) $\frac{1}{3}$

Answers to On Your Own

Section RP 1	Section RP 2	Section RP 3	Section RP 4
1. True	1. False	1. B	1. B
2. False	2. False	2. C	2. B
3. B	3. D		
4. True	4. True		
5. B			
6. C			
7. A			
8. A			
9. Sometines			
10. B			

Section 1.1	Section 1.2	Section 1.3
1. B	1. B	1. True
2. B	2. A	2. C
3. False	3. A	3. Sometimes
	4. D	4. A
	5. B	5. B
	6. False	6. B

Section 1.4	Section 1.5	Section 1.6
1. True	1. Sometimes	1. True
2. B	2. C	2. False
3. D	3. False	3. B
	4. A	4. True
	5. False	5. C

Section 2.1	Section 2.2	Section 2.3
1. False	1. D	1. False
2. True	2. C	2. A
3. D	3. False	3. B
4. B	4. D	4. A
5. B	5. D	
6. True		

Section 2.4	Section 2.5	Section 2.6
1. B	1. False	1. False
2. A	2. C	2. B
3. D	3. D	3. A
4. C	4. A	
5. D	5. True	

Answers to On Your Own

Section 2.7
1. True
2. D
3. B
4. B
5. C

Section 2.8
1. A
2. C

Section 2.9
1. B
2. C
3. D
4. B
5. D

Section 2.10
1. Sometimes
2. A
3. D

Section 3.1
1. A
2. True
3. C
4. True
5. B
6. B

Section 3.2
1. C
2. No
3. A
4. False

Section 3.3
1. A
2. False
3. False
4. B
5. False

Section 3.4
1. False
2. B
3. C
4. B

Section 3.5
1. Sometimes
2. C
3. True
4. D
5. True
6. B

Section 3.6
1. True
2. True
3. False
4. True
5. False
6. True
7. B

Section 3.7
1. C
2. C

Section 3.8
1. B
2. C

Section 3.9
1. False
2. True
3. B
4. B
5. B

Section 3.10
1. False
2. D
3. A
4. False
5. A

Section 4.1
1. B
2. True
3. D
4. B

Section 4.2
1. C
2. A
3. True
4. C

Section 4.3
1. Sometimes
2. Always
3. D
4. B
5. A
6. C
7. D

Answers to On Your Own

Section 4.4
1. D
2. B
3. True
4. D
5. C

Section 4.5
1. D
2. B
3. A
4. C

Section 5.1
1. B
2. B
3. A

Section 5.2
1. A
2. B
3. A

Section 5.3
1. B
2. D

Section 5.4
1. C
2. D

Section 5.5
1. A
2. D
3. C

Section 6.1
1. C
2. D
3. Always
4. D

Section 6.2
1. Sometimes
2. B
3. A
4. C
5. C
6. Sometimes
7. C
8. D
9. D

Section 6.3
1. B
2. False
3. A
4. D
5. A

Section 6.4
1. C
2. B
3. B
4. D
5. C

Section 6.2*
1. B
2. D
3. False
4. C
5. B
6. D

Section 6.3*
1. B
2. C
3. D
4. D

380

Answers to On Your Own

Section 6.4*
1. Sometimes
2. B
3. A
4. C
5. B
6. Sometimes
7. B
8. A
9. D
10. A
11. C

Section 6.5
1. A
2. C

Section 6.6
1. B
2. D
3. B
4. A

Section 6.7
1. False
2. B
3. D
4. C

Section 6.8
1. B
2. A
3. B

Section 7.1
1. A
2. D
3. C

Section 7.2
1. D
2. B

Section 7.3
1. C
2. C
3. B

Section 7.4
1. B
2. D
3. A

Section 7.5
1. B
2. D
3. A

Section 7.6
1. False
2. True
3. False
4. True
5. Sometimes

Section 7.7
1. A
2. C

Section 7.8
1. A
2. B
3. C

Section 7.9
1. False
2. A
3. C
4. D

Section 8.1
1. C
2. C
3. D
4. A
5. A

Section 8.2
1. C
2. B

Section 8.3
1. A
2. C

Answers to On Your Own

Section 8.4
1. B
2. B
3. True

Section 8.5
1. A

Section 8.6
1. True
2. B
3. A

Section 9.1
1. C
2. D
3. A

Section 9.2
1. D
2. C
3. B
4. B

Section 9.3
1. A
2. B

Section 9.4
1. D
2. C
3. B
4. B
5. B

Section 9.5
1. A
2. D
3. C

Section 9.6
1. C
2. B
3. D
4. C

Section 10.1
1. B
2. Sometimes
3. Sometimes
4. Always
5. False
6. A
7. D

Section 10.2
1. False
2. True
3. B
4. False
5. C
6. True

Section 10.3
1. C
2. False
3. No
4. Yes
5. B

Section 10.4
1. Never
2. Yes
3. No
4. Yes
5. A

Section 10.5
1. Yes
2. Yes
3. D
4. False

Section 10.6
1. False
2. True
3. True
4. A
5. B

Section 10.7
1. True
2. False
3. True
4. True

Section 10.8
1. Sometimes
2. True
3. C
4. B
5. C

Section 10.9
1. C
2. B
3. D

Answers to On Your Own

Section 10.10	Section 10.11	Section 10.12
1. C	1. B	1. B
2. C	2. A	2. A
3. A	3. D	
4. True		
5. B		

TO THE OWNER OF THIS BOOK

We hope that you have found *Study Guide, Volume One, for Stewart's Calculus,* Third Edition, useful. So that this book can be improved in a future edition, would you take the time to complete this sheet and return it? Thank you.

School and address: _____

Department: _____

Instructor's name: _____

1. What I like most about this book is: _____

_____

_____

2. What I like least about this book is: _____

_____

_____

3. My general reaction to this book is: _____

_____

4. The name of the course in which I used this book is: _____

_____

5. Were all of the chapters of the book assigned for you to read? _____

   If not, which ones weren't? _____

 6. In the space below, or on a separate sheet of paper, please write specific suggestions for improving this book and anything else you'd care to share about your experience in using the book.

_____

_____

_____

_____

_____

Optional

Your name: _____ Date: _____

May Brooks/Cole quote you, either in promotion for *Study Guide, Volume One, for Stewart's Calculus*, Third Edition, or in future publishing ventures?

Yes: _____ No: _____

Sincerely,

Richard St. Andre

FOLD HERE

FOLD HERE

Performance, reliability, and the most power for your dollar—it's all yours with Maple V!

Maple V Release 3 Student Edition *for Macintosh or DOS/Windows* Just $99.00.

Offering numeric computation, symbolic computation, graphics, and programming, **Maple V Release 3 Student Edition** gives you the power to explore and solve a tremendous range of problems with unsurpassed speed and accuracy. Featuring both 3-D and 2-D graphics and more than 2500 built-in functions, **Release 3** offers all the power and capability you'll need for the entire array of undergraduate courses in mathematics, science, and engineering.

Maple V's *vast* library of functions also provides sophisticated scientific visualization, programming, and document preparation capabilities, **including the ability to output standard mathematical notation.**

With Release 3, you can:

- plot implicit equations in 2-D and 3-D

- generate contour plots

- apply lighting (or shading) models to 3-D plots and assign user-specified colors to each plotted 2-D function

- view 2-D and 3-D graphs interactively and use Release 3's animation capabilities to study time-variant data

- view and print documents with standard mathematical notation for Maple output (including properly placed superscripts, integral and summation signs of typeset quality, matrices, and more)

- save the state (both mathematical and visual) of a Maple session at any point—and later resume work right where you left off

- migrate Maple worksheets easily across platforms. (This is especially valuable for students using Maple V on a workstation in a computer lab who then want to continue work on their own personal computers)

- export to LaTeX and save entire worksheets for inclusion in a publication-quality document *(New!)*

ORDER FORM

Yes! Please send me Maple V Release 3 Student Edition

_____ For Dos/Windows (ISBN: 0-534-25560-4) @ $99

_____ For Macintosh (ISBN: 0-534-25561-2) @ $99

Subtotal _____

(Residents of AL, AZ, CA, CO, CT, FL, GA, IL, IN, KS, KY, LA, MA, MD, MI, MN, MO, NC, NJ, NY, OH, PA, RI, SC, TN, TX, UT, VA, WA, WI must add appropriate sales tax)

Tax _____

Payment Options

Handling _____

_____ Purchase Order enclosed. Please bill me.

_____ Check or Money Order enclosed.

Total _____

_____ Charge my _____ VISA _____ MasterCard _____ American Express

Card Number _____ Expiration Date _____

Signature_____

Please ship to: (Billing and shipping address must be the same.)

Name_____

Department _____ School _____

Street Address _____

City _____ State_____ Zip+4_____

Office phone number (_____) _____

_____ **PLEASE SEND ME SITE-LICENSE INFORMATION!**

You can fax your response to us at 408-375-6414 or e-mail your order to: info@brookscole.com or detach, fold, secure, and mail with payment.

SECURE WITH TAPE

FOLD HERE

**NO POSTAGE
NECESSARY
IF MAILED
IN THE
UNITED STATES**

B U S I N E S S   R E P L Y   M A I L

FIRST CLASS   PERMIT NO. 358   PACIFIC GROVE, CA

POSTAGE WILL BE PAID BY ADDRESSEE

ATTN: _____MARKETING_____

**Brooks/Cole Publishing Company
511 Forest Lodge Road
Pacific Grove, California 93950-9968**

FOLD HERE

Now symbolic computation and mathematical typesetting is as accessible as your Windows™-based word processor

Scientific WorkPlace™ 2.0 Student Edition for Windows

Ideal for homework, projects, term papers, or just writing home—choose from a variety of predesigned styles

"Scientific WorkPlace is a heavy-duty mathematical word processor and typesetting system that is able to expand, simplify, and evaluate conventional mathematical expressions and compose them as elegant printed mathematics. " —Roger Horn, University of Utah

"The thing I like most about Scientific WorkPlace is its basic simplicity and ease of use." —Barbara Osofsky, Rutgers University

Easy access to a powerful computer algebra system
inside **your technical word-processing documents!**

Scientific WorkPlace is a revolutionary program that gives you a "work place" environment—a single place to do all your work. It combines the ease of use of a technical word processor with the typesetting power of TeX and the numerical, symbolic, and graphic computational facilities of the **Maple**® V computer algebra system. All capabilities are included in the program—you don't need to *own* or *learn* TeX, LaTeX, or Maple to use *Scientific WorkPlace*—**everything for super productivity is included in one powerful tool for just $162!**

With *Scientific WorkPlace*, you can enter, solve, and graph mathematical problems right in your word-processing documents in seconds, with no clumsy cut-and-paste from equation editors or clipboards. *Scientific WorkPlace calculates answers* quickly and accurately, then prints your work in professional-quality documents using TeX's internationally accepted mathematical typesetting standard.

Install *Scientific WorkPlace* **and watch your productivity soar!** You'll be creating impressive documents in a fraction of the time you would spend using any other program!

ORDER FORM

Yes! Please send me Scientific Workplace 2.0 Student Edition for Windows

_____ copies (ISBN: 0-534-25597-3) @ $162 each   _____

(Residents of AL, AZ, CA, CO, CT, FL, GA, IL, IN, KS, KY, LA, MA, MD, MI, MN, MO,
NC, NJ, NY, OH, PA, RI, SC, TN, TX, UT, VA, WA, WI must add appropriate sales tax.)   Tax   _____

Payment Options   Handling   _____
_____ Purchase Order enclosed. Please bill me.
_____ Check or Money Order enclosed.   Total   _____
_____ Charge my _____ VISA _____ MasterCard _____ American Express

Card Number _____ Expiration Date _____

Signature_____

Please ship to: (Billing and shipping address must be the same.)

Name_____

Department _____ School _____

Street Address _____

City _____ State_____ Zip+4_____

Office phone number (_____) _____

You can fax your response to us at 408-375-6414 or e-mail your order to: info@brookscole.com
or detach, fold, secure, and mail with payment.

SECURE WITH TAPE

FOLD HERE

BUSINESS REPLY MAIL

FIRST CLASS   PERMIT NO. 358   PACIFIC GROVE, CA

POSTAGE WILL BE PAID BY ADDRESSEE

ATTN: MARKETING

**Brooks/Cole Publishing Company
511 Forest Lodge Road
Pacific Grove, California 93950-9968**

FOLD HERE